EQUITY, EFFICIENCY, AND SOCIAL CHOICE

Equity, Efficiency, and Social Choice

DONALD E. CAMPBELL

CLARENDON PRESS · OXFORD
1992

Oxford University Press, Walton Street, Oxford OX2 6DP
Oxford   New York   Toronto
Delhi   Bombay   Calcutta   Madras   Karachi
Petaling Jaya   Singapore   Hong Kong   Tokyo
Nairobi   Dar es Salaam   Cape Town
Melbourne   Auckland
and associated companies in
Berlin   Ibadan

Oxford is a trade mark of Oxford University Press

Published in the United States
by Oxford University Press, New York

British Library Cataloguing in Publication Data
Data available

Library of Congress Cataloging in Publication Data
Campbell, Donald E.
    Equity, efficiency, and social choice/Donald E. Campbell.
        p.   cm.
    Includes bibliographical references and index.
    1. Social choice. I. Title.
    HB846.8.C36 1992   302'.13 – dc20   91-28812
        ISBN 0-19-828708-9

Typeset by Colset Pte. Ltd., Singapore
Printed in Great Britain by Biddles Ltd, Guildford & King's Lynn

For
Samantha and Jordan

Preface

THIS book represents my recent work on social choice in economic environments. It establishes the classical impossibility theorems under the individual preference restrictions typically assumed for resource allocation models. More important, the book shows that much sharper impossibility theorems are embedded in the topological and algebraic framework of allocation space. These new results reveal there are no efficiency–equity trade-offs if individual behaviour is function of ordinal preference information only.

Each chapter contains a variety of definitions, lemmas, and theorems. They have been numbered seriatim, as they appear. When item i in chapter c is summoned it is cited as c.i, unless the reference occurs in chapter c itself, in which case reference is to item i alone.

It is a pleasure to acknowledge several sources of support. First, I am grateful to the Social Sciences and Humanities Research Council, Canada, for financial support from beginning to end. And I received financial assistance from the National Science Foundation, USA, while preparing the final manuscript.

I acknowledge a very substantial debt to my colleagues for advice along the way. I am particularly grateful to Marcus Berliant, Georges Bordes, Larry Epstein, James Foster, Leonid Hurwicz, Jerry Kelly, Michel Le Breton, Ryo-ichi Nagahisa, Charles Plott, Amartya Sen, Steve Slutsky, William Thomson, John Weymark, and especially to Kenneth Arrow who has been exceptionally encouraging and helpful over the years. I am thankful to Wanda Tidline for her help in preparing the index.

Finally, I acknowledge the support of my wife Diane and my children Samantha and Jordan. What would have been a merely enjoyable academic career has been deeply satisfying as a result of their love and encouragement. But I would find any career rewarding with them at my side.

Contents

Part I
Introduction

Introduction

Part I of this book provides an introduction to the standard terminology and definitions employed in the study of individual and
social choice. It also proves some basic theorems on binary relations
(Chapter 1) and on the topological properties of continuous relations on connected spaces (Chapter 2). Anyone with a grounding in
these fundamentals will want to begin with Part II which presents
the classical social choice theorems in a unified way that renders
them susceptible of generalization and extension to connected
spaces, and to allocation spaces and economic preferences in particular. Part III deals with connected spaces. Continuity of *social*
preference is exploited in a novel way, and much sharper
impossibility theorems emerge. The results on discrete spaces do not
become irrelevant at this point. They are applied to three-element
subspaces and then these local results are 'integrated' over the entire
space, and then strengthened by imposing continuity. Part IV shows
how to apply these stronger impossibility theorems to judiciously
chosen subspaces of the allocation space and associated economic
preferences in a way that leads to proofs of the theorems for the
whole allocation framework. The sharper theorems have very
precise implications for the nature of efficiency–equity trade-offs.
Essentially, they reveal that there are no meaningful efficiency–equity trade-offs if Arrow's independence axiom — our term
for Independence of Irrelevant Alternatives — is imposed. The
penultimate chapter shows that this controversial condition is
actually implied by a very mild incentive compatibility condition.
The final chapter offers some concluding remarks.

1   Binary Relations

WE begin with an examination of the conditions under which choice from an arbitrary set of alternatives can be derived from choice over two-element sets. The universal set of alternatives is some non-empty set X, which is determined in advance. The properties of choice from two-element sets are captured by means of a *binary relation* on X, which is some collection of ordered pairs of members of X.

1. Binary relation. A binary relation R on X is a subset of $X \times X$.

The statement $(x,y) \in R$ means that the individual would not object to being presented with alternative x when only x and y are available. The more suggestive notation xRy is usually employed to indicate that (x,y) belongs to R, but the subset representation is often more useful in presenting examples and theorems. We will take advantage of both. If xRy holds but yRx does not then the individual will always choose x when only x and y are available. In that case we say that x is strictly preferred to y. If xRy and yRx both hold then the individual is said to be indifferent between x and y and would not mind if the choice from $\{x,y\}$ were determined by an independent agent.

2. Reflexivity. A binary relation R is reflexive if xRx holds for all $x \in X$.

Although it is not necessary to determine which alternative in the set $\{x,y\}$ the individual would choose in case $x = y$, we do learn something when we discover that R is reflexive. It makes no sense to say that x is strictly preferred to itself, and therefore a reflexive relation cannot represent strict preference. It must represent weak preference in the sense that xRy reveals that x is at least as desirable as y; it *may* not be more desirable than y, as in the case $y = x$. A condition that implies reflexivity is *completeness*.

3. Complete relation. A binary relation R is complete if for all $x,y \in X$ either xRy or yRx holds.

If R is complete and $x = y$ then xRx must hold, and hence R is reflexive, because completeness implies that either xRx or xRx must hold. If R is complete and xRy holds we say that x is preferred or indifferent to y; x *may* be strictly preferred to y. We cannot tell from the observation xRy alone because x may be indifferent to y, as in the case $x = y$. It is possible to determine whether x is strictly preferred to y by examining R further. If xRy holds but yRx does *not* hold then x is at least as desirable as y but y is not at least as desirable as y. Then we can say that x is strictly preferred to y.

If R is *asymmetric* we interpret it as strict preference. We usually let P denote a strict preference relation.

4. Asymmetry. The binary relation P is asymmetric if for all $x,y \in X$, xPy implies that yPx is false.

Because xPy and yPx can never be true simultaneously when P is asymmetric we can conclude that an asymmetric relation represents strict preference; it is absurd to have x strictly preferred to y and y strictly preferred to x, and this is exactly what is ruled out by asymmetry.

The strict preference part, or *asymmetric factor*, AR of any binary relation R can be rendered easily.

5. Asymmetric factor. $AR = \{ (x,y) \in R \mid (y,x) \notin R \}$.

It is easy to prove that $AP = P$ for any asymmetric relation P. We can also use the basic relation R to determine the indifference relation, or the *symmetric factor* of R, denoted SR.

6. Symmetric factor. $SR = \{ (x,y) \in R \mid (y,x) \in R \}$.

The *null relation* is defined to be the complete relation R^0 on X for which $SR^0 = R^0$ holds. Clearly, $R^0 = \{ (x,y) \mid x,y \in X \}$. It is a special case of a *symmetric* binary relation.

7. Symmetry. R is symmetric if $R = SR$.

Note that a symmetric relation need not be complete. Obviously, SR itself is symmetric for any R.

8. Example. $X = \{ w,x,y,z \}$ and
$$R = \{ (w,x),(w,y),(w,z),(x,y),(x,z),(y,z),$$
$$(z,y),(w,w),(x,x),(y,y),(z,z) \}.$$
Then R is complete (and reflexive). It is not asymmetric because

we have both yRz and zRy; and it is not symmetric because wRx holds although xRw does not. We have $AR = \{ (w,x),(w,y), (w,z),(x,y),(x,z) \}$ and $SR = \{ (y,z),(z,y),(w,w),(x,x),(y,y), (z,z) \}$.

If the symbols R, P, and I are used without qualification it will be understood that R is the primitive relation, P is its asymmetric factor, and I is its symmetric factor. The following elementary property of complete relations will be used repeatedly and without specific reference.

9. Proposition. If R is a complete relation on X then for all $x,y \in X$ exactly one of the following holds: xPy, yPx, xIy.

Proof. Suppose xRy. Then yPx cannot hold. If yRx then xIy but not xPy. If yRx is false then xPy but not xIy. Suppose that xRy is false. Then xPy and xIy are both false. But yRx because R is complete, so yPx holds. ■

Obviously, there are situations in which choices are made from sets containing more than two members. Such decisions are represented by means of a *choice function*.

10. Choice function. A choice function is any function C mapping 2^X, the family of subsets of X, into itself in such a way that $C(Y) \subset Y$ holds for all $Y \subset X$. Of special interest is max$R$, the choice function generated by a binary relation R.

11. Definition. The choice function max$R$ generated by the binary relation R on X is defined by setting max$R(Y) = \{x \in Y \mid (y,x) \in AR$ implies $y \notin Y \}$.

If x belongs to max$R(Y)$ we say that it is an *R-maximal* element of Y. If the binary relation R is clearly understood we will simply say that x is maximal in Y.

Only complete relations are used in our discussion of social choice. If R is complete then max$R(Y) = \{x \in Y \mid xRy \, \forall Y \in Y\}$. In general, $\{x \in Y \mid xRy \, \forall y \in Y\} \subset$ max$R(Y)$. And if R is complete and $x \in$ max$R(Y)$ then for each $y \in X$ either xRy or yRx holds. Therefore, by Definition 11, either xRy alone holds or both xRy and yRx hold, for each $y \in Y$. Thus max$R(Y) \subset \{x \in Y \mid xRy \, \forall y \in Y\}$ if R is complete, and hence max$R(Y) = \{x \in Y \mid xRy \, \forall y \in Y\}$ if R is complete. If P is asymmetric then $AP = P$ and thus max$P(Y) = \{x \in Y \mid yPx$ implies $y \notin Y \}$ for any asymmetric relation P.

Now we consider how choice from a three-element set can be based on choice in two-element subsets. Consider any three alternatives x, y, and z. Suppose that P is asymmetric and xPy, yPz, and zPx. Then the decision-taker would choose x from $\{x,y\}$, y from $\{y,z\}$, and z from $\{x,z\}$. In this case the choice on the three-element set $Y = \{x,y,z\}$ cannot be inferred from binary choice: x is not best in Y because z is strictly preferred to x, z is not best in y because Y is strictly preferred to z, and y is not best in Y because x is strictly preferred to y. One way to rule out this possibility is to require xPz to hold whenever xPy and yPz both hold. In that case P is said to be transitive.

12. Transitivity. A binary relation R on X is transitive if for all x,y,z in X, xRy and yRz imply xRz.

13. Proposition. If R is transitive and $x_1,x_2, \ldots ,x_m$ is a finite list of alternatives in X then $x_1Rx_2R \ldots x_{m-1}Rx_m$ implies x_1Rx_m.
Proof. We have x_1Rx_3 because R is transitive. Because x_1Rx_3,x_3Rx_4, and R is transitive we have x_1Rx_4. Similarly, x_1Rx_5,x_1Rx_6, and so on. In general, x_1Rx_m. ∎
If we have a finite list $x_1,x_2, \ldots ,x_m$ of alternatives and x_iPx_{i+1} for $i = 1,2, \ldots ,m-1$ we will usually denote this by writing $x_1Px_2P \ldots x_{m-1}Px_m$. But this is not meant to imply that x_iPx_j necessarily holds whenever $i < j$. This *will* be the case if P is transitive. (To prove this apply Proposition 13 to the list $x_i,x_{i+1}, \ldots ,x_j$.) Therefore, transitivity imposes enough structure on a binary relation to guarantee the existence of at least one maximal element in any non-empty finite set. This will be proved following another example and an elementary theorem. The example demonstrates that a relation need not be transitive even if its asymmetric factor is.

14. Example. $X = \{x,y,z\}$ with $x \neq y \neq z \neq x$ and
$$R = \{ (x,y),(x,z),(z,x),(y,z),(z,y),(x,x),(y,y),(z,z) \}.$$
$P = \{ (x,y) \}$, the asymmetric factor of R, is vacuously transitive: the situation $aPbPc$ can never arise for $a,b,c \in X$. But R is not transitive because we have yRz and zRx although yRx does not hold. The asymmetric factor of a transitive relation *is* always transitive, as we now show.

15. Proposition. If a relation R is transitive then its asymmetric factor P is transitive.

Proof. Suppose R is transitive and xPy and yPz hold. Then xRy and yRz by definition, and thus xRz because R is transitive. Suppose zRx. But zRx and xRy imply zRy by transitivity of R. This contradicts yPz. Thus, zRx cannot hold. We have shown that xRz holds and zRx does not. Then xPz and P is transitive. ∎

Now we prove our first theorem on the existence of maximal elements.

16. Proposition. If the relation R is transitive then for any finite and non-empty subset Y of X the choice set $\max R(Y)$ is not empty.

Proof. Set $Y = \{y_1, y_2, \ldots, y_m\}$ and let P be the asymmetric factor of R. Define $x_1, x_2, \ldots$ inductively. Set $x_1 = y_1$. If $y_i P x_t$ for some $y_i \in Y - \{x_1, x_2, \ldots, x_t\}$ set $x_{t+1} = y_i$ for the smallest integer i such that $y_i P x_t$ and $y_i \in Y - \{x_1, \ldots, x_t\}$. If there is no such y_i terminate the sequence $x_i, x_2, \ldots, x_t$ with term t. Suppose the sequence terminates at iteration n. We show that $x_n \in \max R(Y)$. If $y_i P x_n$ for some $y_i \in Y$ then $y_i = x_j$ for some $j < n - 1$. But then $x_j P x_n P_{n-1} P \ldots x_{j+1} P x_j$. P is transitive by Proposition 15. Therefore, $x_j P x_j$ by Proposition 13, a contradiction. Therefore, $y_i P x_n$ does not hold for any $y_i \in Y$ and hence $x_n \in \max R(Y)$. Therefore, $\max R(Y) \neq \varnothing$ for any finite set Y. ∎

We know that AR may be transitive even if R is not (Example 14) but AR is always transitive if R is (Proposition 15). Therefore, transitivity of the asymmetric factor is weaker than transitivity of the relation itself. Proposition 16 shows that either suffices for the existence of maximal elements in finite sets because $\max AR(Y) = \max R(Y)$ in general. Therefore, transitivity of the asymmetric factor is a more general sufficient condition for the existence of a maximal element. But even this is unnecessarily strong. To demonstrate this we introduce a new condition, *acyclicity*. It is not necessary to assume that xPy and yPz imply xPz in order to guarantee that $Y = \{x, y, z\}$ contains a P-maximal element. As long as zPx is false when xPy and yPz both hold then bPx will not hold for any $b \in \{x, y, z\}$. Acyclicity rules out zPx whenever xPy and yPz. In order to ensure that *any* finite set contains a best element we will have to consider chains $aPbPc \ldots PyPz$ of more than three alternatives.

17. Acyclicity. The binary relation R is acyclic if for $P = AR$ and any finite collection $x_1, x_2, \ldots, x_t \in X$ we have $x_1 \neq x_t$ if $x_1 P x_2 P \ldots x_{t-1} P x_t$.

18. Proposition. $\max R(Y) \neq \varnothing$ for all finite and non-empty subsets Y of X if and only if R is acyclic.

Proof. Let P be the asymmetric factor of R.

(i) If R is not acyclic there is a list $x_1, x_2, \ldots, x_m$ of alternatives in X such that $x_1 P x_2 P \ldots x_{m-1} P x_m P x_1$. Set $Y = \{x_1, x_2, \ldots, x_m\}$. Then $\max R(Y) = \varnothing$ because $x_{t-1} P x_t$ holds for all $t > 1$ and thus $x_t \notin \max R(Y)$, and also $x_1 \notin \max R(Y)$ because $x_m P x_1$. Therefore, if $\max R(Y) \neq \varnothing$ for all finite Y then R is acyclic.

(ii) Suppose that R is acyclic. Define the asymmetric binary relation Q on X:

> $\forall \, x, y \in X$, xQy if and only if xPy or there is some finite list of alternatives $a_1, a_2, \ldots, a_m$ in X such that $xPa_1 Pa_2 P \ldots a_{m-1} Pa_m Py$.

Q is asymmetric because xQy and yQx imply that $xPa_1 Pa_2 \ldots Pa_m PyPb_1 Pb_2 \ldots Pb_n Px$ holds for appropriate choices of a_i and b_j and this contradicts acyclicity of P. Because Q is asymmetric $\max Q(Y) = \{x \in Y \mid yQx \text{ implies } y \notin Y\}$ for finite Y. But Q is transitive: xQy and yQz implies $xPa_1 Pa_2 \ldots Pa_m PyPb_1 Pb_2 \ldots Pb_n Pz$ for appropriate choice of a_i and b_j and therefore xQz by definition. Therefore, $\max Q(Y) \neq \varnothing$ by Proposition 16. But $\max Q(Y) \subset \max P(Y)$. If yPx then yQx by definition, and thus $\max Q(Y) \subset \max P(Y)$. Therefore, $\varnothing \neq \max Q(Y) \subset \max P(Y) = \max R(Y)$. ∎

Part (ii) of Proposition 18 can be proved directly by employing the proof of Proposition 16; when we get to $x_j P x_{j+1} \ldots P x_j$ we have a violation of acyclicity. Proposition 18 is stronger than Proposition 16 because every transitive relation is acyclic. Let P be the asymmetric factor of the transitive relation R. If $x_1 P x_2 \ldots P x_m$ then $x_1 P x_m$ by Proposition 13. Then $x_m P x_1$ is false because P is asymmetric. Hence P and R are acyclic. But there are acyclic relations that are not transitive.

19. Example. $X = \{x, y, z\}$ with $x \neq y \neq z \neq x$ and $P = \{(x, y), (y, z)\}$.

P is not transitive because xPz is false, even though xPy and yPz.

But P is acyclic because zPx does not hold.

Although acyclicity is necessary and sufficient for the existence of a best (or maximal) element in every finite set, the stronger assumption of transitivity is often employed because of its finer structure. In particular, only chains with two links, such as $xRyRz$, need be examined in order to test for transitivity. Of particular importance are binary relations that are complete and transitive, and relations that are asymmetric and transitive. These relations are used repeatedly and are referred to as preorders and quasiorders, respectively.

20. Preorder. A binary relation is called a preorder if it is complete and transitive.

21. Quasiorder. A binary relation is called a quasiorder if its asymmetric factor is transitive. A complete quasiorder is said to have the quasitransitivity property.

The next proposition is a convenient characterization of preorders.

22. Proposition. A complete relation is transitive if and only if its asymmetric and symmetric factors are both transitive.

Proof. Let R be a complete relation and let P and I be the asymmetric and symmetric factors, respectively.

(i) Suppose that R is transitive. Then P is transitive by Proposition 15. If xIy and yIz then, by definition, xRy, yRx, yRz, and zRy. Then xRz because R is transitive and xRy and yRz both hold. Also, zRx, because zRy and yRx. Therefore, xIz and I is transitive.

(ii) Suppose P and I are both transitive. If xRy and yRz but xRz does not hold then zPx because R is complete. If xPy we have zPy, by transitivity of P, contradicting yRz. Therefore, xIy must hold. If yIz then xIz, by transitivity of I, contradicting zPx. Therefore, yPz. But yPz and zPx imply yPx, by transitivity of P, contradicting xRy. Therefore xRz must hold after all and hence R is transitive. ∎

Part (i) of the proof does not depend upon the completeness of R. But completeness is essential for part (ii), as the next example reveals.

23. Example. $X = \{x,y,z\}$ with $x \neq y \neq z \neq x$, and
$$R = \{(x,y),(y,z),(z,y),(x,x),(y,y,),(z,z)\}.$$

R is not complete because neither xRz nor zRx is true. R is intransitive because xRy and yRz both hold but xRz does not. But P and I are both transitive as one can easily verify: $P = \{(x,y)\}$ and $I = \{(y,z),(z,y),(x,x)(y,y),(z,z)\}$.

A *linear order* is a preorder for which indifference is ruled out unless the alternatives in question are identical.

24. Linear order. R is a linear order if it is a preorder and, $\forall\, x,y \in X$, xRy and yRx imply $x = y$.

A linear order R is characterized by the following three properties:

24a $\forall\, x \in X$, xRx.
24b If $x,y, \in X$ and $x \neq y$ then either xRy or yRx holds but not both.
24c R is transitive.

If P is the asymmetric factor of a linear order R then P satisfies 24b and 24c. Therefore, we will also use the term linear order to apply to an asymmetric relation R satisfying 24b and 24c.

25. Example. Let $\mathbf{E}^k$ denote k dimensional Euclidean space.

$\forall\, x,y \in \mathbf{E}^k$, $x \geq y$ means that $x_i \geq y_i$ holds for $i = 1,2,\ldots,k$. Define R by setting xRy if and only if $y \geq x$ implies $y = x$.

If $k = 1$ then $\geq$ is a reflexive linear order and the relations R and $\geq$ are equivalent. For any k, $x \geq y$ and $y \geq x$ imply $x = y$. If $k = 1$ and $x \neq y$ either $x \geq y$ or $y \geq x$ holds but not both. If $k \geq 2$, $x = (1,2,\ldots)$, and $y = (2,1,\ldots)$ then neither $x \geq y$ nor $y \geq x$ holds, so $\geq$ is not complete if $k \geq 2$. Note that $\geq$ is transitive for all k; that is, $\geq$ is transitive for all k but it is complete if and only if $k = 1$. On the other hand, the relation R is complete for all k. If xRy does not hold then $y \geq x \neq y$ so yRx holds. But 24b and 24c fail to hold for $k \geq 2$. Choose $x,y,z \in E^k$ such that $x_i = y_i = z_i$ for $i > 2$ and $(x_1,x_2) = (1,2)$, $(y_1,y_2) = (4,1)$, $(z_1,z_2) = (3,3)$. Then xRy and yRx although $x \neq y$ so 24b fails. xRy and yRz but xRz does not hold and 24c fails.

The advantage of a linear order is that its choice set, $\max R(Y)$, contains one and only one alternative for all finite, non-empty Y, as we now prove.

26. Proposition. If R is reflexive then $\max R(Y)$ is a singleton for every finite, non-empty $Y \subset X$ if and only if R is a linear order.

Proof. (i) Suppose that max$R(Y)$ is a singleton for all finite, non-empty $Y \subset X$. Let P be the asymmetric factor of R.

If $x \neq y$ but neither xRy nor yRx holds then max$R(\{x,y\}) = \{x,y\}$, contradicting the fact that max$R(Y)$ is a singleton for each finite and non-empty Y. If $x \neq y$ and both xRy and yRx hold then again we get max$R(\{x,y\}) = \{x,y\}$. Therefore, R is complete if it is reflexive, and hence it is a linear order if P is transitive. Suppose $xPyPz$. Then $x \neq z$. If zPx then max$R(\{x,y,z\}) = \varnothing$, not a singleton set. Therefore zPx cannot hold and hence xRz must hold. We have xPz because xRz and zRx cannot both hold. Therefore P is transitive.

(ii) Suppose that R is a linear order. max$R(Y) \neq \varnothing$ for all finite non-empty Y by Proposition 16. If $x \in$ max$R(Y)$ and $y \in Y - \{x\}$ then yPx does not hold because $x \in$ max$R(Y)$, and therefore xPy must hold because $y \neq x$ and R is a linear order. Therefore, xPy if $x \in$ max$R(Y)$ and $y \in Y - \{x\}$. Thus max$R(Y) = \{x\}$. Each choice set is a singleton. ∎

The next proposition establishes a useful property of preorders.

27. Proposition. Let R be a preorder with asymmetric and symmetric factors P and I respectively. Then $\forall\, x,y,z \in X$, xPz holds if $xRyRz$ and either xPy or yPz.

Proof. Suppose xPy and yIz but xPz does not hold. Then zRx because R is complete. And yRz by definition of I. Then yRx by transitivity, contradicting xPy. Therefore, xPz must hold.

Suppose xRy and yPz. If zRx then zRy by transitivity, contradicting yPz. Therefore, xRz because R is complete, and hence xPz. ∎

In words, if x is at least as desirable as y and y is at least as desirable as z then x is strictly preferred to z unless x, y, and z are indifferent to each other. This will not necessarily be the case if R is not a preorder as the following example shows.

28. Example. $X = \{x,y,z\}$ with $x \neq y \neq z \neq x$.
$R = \{\, (x,y), (y,z), (z,y), (x,z), (z,x)\, (x,x)\, (y,y),$
$\quad (z,z)\, \}$.

R is a complete binary relation but it is not transitive because yRx does not hold although yRz and zRx do. We have xPy and yIz but xPz is false.

The next proposition characterizes preorders in terms of the choice sets generated by them.

29. Proposition. If R is a preorder and $\max R(Y) \neq \emptyset$ then $\max R(Y)$ is the only non-empty subset Z of Y such that for all $x \in Z$ we have xIy if $y \in Z$ and xPy if $y \in Y - Z$.

Proof. Set $Z = \max R(Y)$. Then $x,y \in Z$ implies xRy and yRx and hence xIy. If $x \in Z$ and $y \in Y - Z$ then zPy for some $z \in Y$. But $x \in Z$ so zPx cannot hold. Thus xRz because R is complete. Then $xRzPy$ and hence xPy by Proposition 27. Then $\max R(Y)$ has the stated property. If Z' also has this property we have xRy for all $x \in Z'$ and $y \in Y$, and thus $Z' \subset Z$. If $x \in Z - Z'$ and $y \in Z'$ then xIy because $x,y \in Z$. Then yPx cannot hold, although $y \in Z'$ and $x \notin Z'$. Therefore, Z' does not have the stated property unless $Z' = Z$. ∎

For any binary relation R we can define the 'preferred set' Rx and the 'less desirable set' xR for arbitrary $x \in X$.

30. Definition. $\forall x \in X$, $Rx = \{y \in X \mid yRx\}$, and $xR = \{y \in X \mid xRy\}$.

This notation leads to the following useful characterization of transitivity for any reflexive and symmetric relation (such as the symmetric factor of any complete binary relation).

31. Proposition. If I is a reflexive and symmetric binary relation then it is transitive if and only if $\forall x,y \in X, Ix \cap Iy \neq \emptyset$ implies $Ix = Iy$.

Proof. (i) Suppose that I is symmetric and transitive. If $z \in Ix \cap Iy$ then zIx and zIy. And xIz because I is symmetric. If $a \in Ix$ then $aIxIz$ and thus aIz by transitivity of I. Then $aIzIy$ and thus aIy by transitivity of I. Therefore, $Ix \subset Iy$. Similarly, $Iy \subset Ix$.

(ii) Suppose that I is symmetric and reflexive and $\forall x,y \in X$, $Ix \cap Iy \neq \emptyset$ implies $Ix = Iy$. Suppose that xIy and yIz. Because I is symmetric, yIx holds, so $y \in Ix \cap Iz$ and thus $Ix = Iz$. But $x \in Ix$ because I is reflexive. Thus, xIz and I is transitive. ∎

32. Equivalence relation. An equivalence relation is a reflexive, symmetric, and transitive binary relation. If R is an equivalence relation then Rx is called an equivalence class. Let $E(R) = \{Rx \mid x \in X\}$ denote the family of equivalence classes.

If R is an equivalence relation then each x in X belongs to one and only one equivalence class. (We have $x \in Rx$. And $x \in Ry \cap Rz$ implies $Ry = Rz$ by Proposition 31.)

If R is symmetric but not reflexive then transitivity of R does not necessarily follow from the statement '$\forall x,y \in X, Rx \cap Ry \neq \emptyset$ implies $Rx = Ry$' as the next example shows.

33. Example. $X = \{x,y,z\}$ where $x \neq y \neq z \neq x$ and $R = \{(x,y), (y,x), (y,z), (z,y)\}$.

We have xRy and yRz but not xRz. Yet R is symmetric. And $Rz = \{y\}$, $Rx = \{y\}$, and $Ry = \{x,z\}$ so either $Ra \cap Rb = \emptyset$ or $Ra = Rb$.

The next proposition is a summary of many of the preceding points.

34. Proposition. The following four statements apply to a reflexive binary relation R on X with asymmetric factor P. ψ is the family of all non-empty, finite subsets of X.

34a $\max R(Y)$ is a singleton for any $Y \in \psi$.

34b $\max R(Y)$ is not empty for any $Y \in \psi$, and for all $x \in \max R(Y)$ and $y \in Y - \max R(Y)$ we have xPy.

34c $\forall Y \in \psi$ and $\forall y \in Y - \max R(Y)$ we have xPy for some $x \in \max R(Y)$.

34d $\max R(Y) \neq \emptyset \ \forall Y \in \psi$.

Then 34a implies 34b implies 34c implies 34d but the converse does not hold for any of these implications.

Proof. If 34a holds then R is a linear order by Proposition 26. Therefore, 34b holds as well. Obviously, 34b implies 34c. And 34c implies that $\max R(Y) \neq \emptyset$ if Y is non-empty and finite so 34c implies 34d. Note that 34b holds if and only if there is some preorder R' such that $AR' \subset R \subset R'$ (Proposition 29), and 34c holds if and only if $AR = AQ$ for some quasiorder Q. By Proposition 18, 34d is equivalent to acyclicity of R.

34d does not imply 34c. Consider Example 19. We have $\max P(X) = \{x\}$ but xPz does not hold so 34c fails. But P is acyclic so 34d holds by Proposition 18 or by direct computation. 34c does not imply 34b: Set $X = \{w,x,y,z\}$ and $P = \{(x,y),(x,z),(y,z)\}$. Then $\max P(X) = \{w,x\}$ but wPy does not hold and 34b fails. It is easy to verify 34c. 34b does not imply 34a. Let $X = \{x,y,z\}$ and $R = X \times X$. Then $\max R(Y) = Y$ for $Y \in \psi$ so 34b holds; 34a obviously fails. ∎

Now define the inverse of a binary relation, which will be useful in streamlining the definitions and proofs of subsequent chapters.

35. Inverse. The inverse of a binary relation R on X is the relation $-R = \{(x,y) \in X \times X \mid (y,x) \in R\}$. Note that $SR = R \cap -R$.

Our last task in this chapter is to show that on any set X a wide variety of linear orders may be defined, even if X is infinite. This is required in Chapter 7, but the other chapters use this property only for finite X. If X is infinite, the proof of the existence of the linear orders rests on the *axiom of choice*, which asserts that if Λ is any index set and X_λ is a non-empty set for each $\lambda \in \Lambda$ then there is a function γ on Λ such that $\gamma(\lambda) \in X_\lambda$ for each $\lambda \in \Lambda$. Actually, we will use *Zorn's lemma* which is equivalent to the axiom of choice (see Kelley, 1955: 33). First we define a *chain* in X given a binary relation $>$ on X. Y is a chain if it is a subset of X and $>$ is a complete linear order on Y. An *upper bound* for $Y \subset X$ with respect to $>$ is a member x of X such that $x > y$ holds for all $y \in Y$.

36. Zorn's lemma. Let X be any set and let $>$ be any quasiorder (21) on X. If each chain in X has an upper bound then X has a maximal element with respect to $>$.

Zorn's lemma allows us to prove that if x and y are any distinct members of X then there is a complete linear order R on X such that $(x,y) \in R$. We actually prove a substantially stronger result. (If X is a finite set of three or more members a proof of the following result can be fashioned easily, without Zorn's lemma or any of its equivalent forms, by setting $X = \{x_1, x_2, \ldots, x_n\}$.)

37. Proposition. Let X be any set and let W, Y, and Z be any three non-empty and mutually disjoint subsets of X. Then there is a complete linear order R on X such that (w,y) and (y,z) belong to R for any $w \in W$, $y \in Y$, and $z \in Z$.

Proof: Let Σ denote the family of *preorders* R on X such that

37a $(w,y), (y,z) \in AR$ whenever $w \in W$, $y \in Y$, and $z \in Z$.

To show that Σ is not empty define the preorder R on X by setting aRb if and only if (1) $a \in Y$ implies $b \notin W$ and (2) $a \notin W \cup Y$ implies $b \notin W \cup Y$. Then R satisfies 37a. Verify that R is a preorder.

Now define the binary relation $>$ on Σ: Set $R > R'$ if and only if $AR' \subset AR$. Note that $R > R'$ implies $SR \subset SR'$ because R' is complete. Then $R > R' > R$ implies $R = R'$. Let Γ be a chain in Σ with respect to $>$. Now we prove

37b   If $(x,y) \in AR$ and $R \in \Gamma$ then $(x,y) \in R'$ for all $R' \in \Gamma$.

Suppose $(x,y) \in AR$ and $R \in \Gamma$. If $R' > R$ and $R' \in \Gamma$ then $(x,y) \in AR'$ by definition of $>$ and thus $(x,y) \in R'$. If $R > R'$ and $R' \in \Gamma$ then $(x,y) \notin R'$ implies $(y,x) \in AR'$ because R' is complete, contradicting the definition of $>$. This proves 37b.

Set $R^* = \cap \{R \mid R \in \Gamma\}$. We have

37c $R^* > R, \forall R \in \Gamma$.

If $(x,y) \in AR$ and $R \in \Gamma$ then $(x,y) \in R^*$ by 37b. And $(y,x) \notin R^*$ because $(y,x) \notin R$. Thus $(x,y) \in AR^*$.

R^* is a complete relation on X because $(x,y) \notin R^*$ implies $(y,x) \in AR$ for some $R \in \Gamma$ and hence $(y,x) \in R^*$ by 37c. If (x,y) and (y,z) both belong to R^* then $(x,y),(y,z) \in R \; \forall R \in \Gamma$ and thus $(x,z) \in R$ for each $R \in \Gamma$ because each is a preorder. Therefore $(x,z) \in R^*$. Therefore, R^* is complete and transitive. Because each R in Σ satisfies 37a the preorder R^* satisfies 37a by 37c. Therefore, $R^* \in \Sigma$. Therefore, R^* is an upper bound for Γ, so Σ has a maximal element Q by Zorn's lemma. We conclude by showing that Q is a linear order.

Suppose $x,y \in X$, $x \neq y$, and $(x,y) \in SQ$. Define Q' by setting $aQ'b$ if and only if (1) aQb and (2) $a = y \neq b$ implies $(y,b) \in AQ$. It is easy to confirm that Q' is a preorder on X satisfying 37a. Obviously $Q' > Q$ but not $Q > Q'$. This contradicts the assertion that Q is maximal in Σ with respect to $>$. Therefore, Q must be a linear order. ∎

2 Topology and Preference

THE aim of social choice theory is to settle public policy disputes, or at least to influence them. Therefore, practical considerations underlie the construction of a model and the formulation of ethical desiderata. In this chapter we come to terms with the fact that neither alternatives nor individual preferences can be specified with complete accuracy. As a result, we require a model that is not unduly sensitive to very slight changes in the outcome. In plain terms, if x is judged to be socially superior to y then anything extremely close to x should be superior to anything extremely close to y. This in turn implies that individual preferences have the same property. Suppose, for instance, that x is socially superior to y because every individual in society strictly prefers x to y. Suppose also that, apart from x and y, everyone prefers any alternative near y to any alternative near x. Then the collective choice process will be unavoidably error prone. Because the alternatives cannot be specified with perfect precision or certainty and x' should prevail against y' if $x' = x$ and $y' = y$ but not otherwise, it will not be possible to use the social ranking to settle a dispute between x and y. To make this point even more forcefully, suppose that alternative x refers to the construction of a bridge and y refers to the status quo. Let Z denote the set of bridges with a weight that is some rational number between 1,000 and 1,100 (tons) and let Z' denote all other bridges weighing between 1,000 and 1,100 tons. Suppose that everyone prefers x to y if x is in Z but everyone prefers y to a bridge in Z'. Then the community cannot be sure of making the correct decision even though one member of any pair of options will have unanimous support.

Of course, the individual preference scheme just described is too bizarre to be taken seriously. We can safely *assume away* any such individual preference relation. And we want to include in the set of axioms that a social decision procedure must satisfy the *requirement* that there be no social preference reversals resulting from a practically insignificant variation in the outcome. Technically, we want *continuous* individual and social preferences, although Kelly (1971) demonstrates that continuity of social preference is quite demanding by showing how stringent are the conditions for continuity of the simple majority-rule relation, even when transitivity is guaranteed. Nevertheless, we claim that it is a natural requirement and will insist on it in Parts III and IV. Continuity of preference is a generalization of the idea of a continuous function. The definition depends on an understanding of *topology*, which we briefly review.

1. Topology. X is a set. A topology for X is a family $\mathfrak{I}$ of subsets of X satisfying:

1a $\varnothing \in \mathfrak{I}$ and $X \in \mathfrak{I}$.

1b $Y \cap Z \in \mathfrak{I}$ if Y and Z belong to $\mathfrak{I}$.

1c   The union of *any* subset of $\mathfrak{I}$ belongs to $\mathfrak{I}$.

A set is said to be *open* if it belongs to $\mathfrak{I}$ and closed if its complement belongs to $\mathfrak{I}$. $\mathfrak{I}$ is the *discrete topology* if it includes every subset of X. The pair $(X, \mathfrak{I})$ is called a topological space. For convenience we often refer to X as a topological space, with $\mathfrak{I}$ implicit.

Suppose that $Y \subset X$ and for each $y \in Y$ there is an open set $N(y) \subset Y$ that contains y. Then $Y = \cup \{N(y) \mid y \in Y\}$ and therefore Y is open by property 1c. On the other hand, if Y is open we can set $N(y) = Y$ for each $y \in Y$. We have $Y = \cup \{N(y) \mid y \in Y\}$ with each $N(y)$ open. Therefore, a set Y is open if and only if every member of Y belongs to some open set that is itself a subset of Y. Therefore, Z is closed if and only if each member of $X - Z$ belongs to some open set that does not intersect Z. If $V \subset X$ and $x \in X$ we say that x is a *closure point* of V if every open set containing x has a non-empty intersection with V. Therefore, a set Z is closed if and only if it contains all of its closure points. The *closure* of arbitrary $V \subset X$ is the set clos V comprised of V and all its closure points. Then V is closed if and only if $V = \text{clos} V$.

A topological space is called a T_1-space if for every $x \in X$ and every $y \neq x$ in X there is an open set $N(x)$ containing x but not y. It is a *Hausdorff* space if for any distinct points x and y in X there are disjoint open sets $N(x)$ and $N(y)$ containing x and y respectively. The space is *normal* if for every two closed subsets F_1 and F_2 of X such that $F_1 \cap F_2 = \varnothing$ there are two open sets Y_1 and Y_2 such that $F_1 \subset Y_1$, $F_2 \subset Y_2$, and $Y_1 \cap Y_2 = \varnothing$. Note that $\{x\}$ is closed if x is a point in a T_1-space: If $z \neq x$ there is some open set Y_z

containing z such that $x \notin Y_z$. By 1c, $Y = \cup \{ Y_z \,|\, z \in X \text{ and } z \neq x \}$ is open and thus $\{x\}$, the complement of Y, is closed. Therefore, every normal T_1 space is Hausdorff. Obviously, every Hausdorff space is T_1. (The Hausdorff property is usually included in the definition of normality. For example, see Dugundji (1966: 144). But we do not require this for the proof of Urysohn's theorem below, which is the only implication of normality that is exploited in this book.) A *connected* topological space is one that cannot be partitioned into disjoint, non-empty, open subsets. That is, X is connected if one cannot find open sets Y, $Z \in \mathfrak{I}$ such that $X = Y \cup Z$, $Y \cap Z = \varnothing$, and $Y \neq \varnothing \neq Z$.

2. Relative topology. If $(X, \mathfrak{I})$ is a topological space and Y is a subset of X the relative topology for Y is $\mathfrak{I}_Y = \{ Y \cap Z \,|\, Z \in \mathfrak{I} \}$. (Exercise: Prove that $\mathfrak{I}_Y$ is a topology.)

The topology entails the idea of closeness. The point $x \in X$ is supposed to be close to the set $Y \subset X$ if x is a closure point of Y. This allows us to define continuous functions and continuous preference relations. If x is a closure point of Y then continuity of a function g on X requires that $g(x)$ be close to $\{ g(y) \,|\, y \in Y \}$. Intuitively, as the open sets N containing x get smaller and smaller the points in $N \cap Y$ are getting closer and closer to x, in terms of the concept of closeness embodied in $\mathfrak{I}$, and $g(N \cap Y)$ must approach $g(x)$ for g to be continuous. Formally, continuity of the function g mapping topological space $(X, \mathfrak{I})$ into topological space $(X', \mathfrak{I}')$ is defined as follows: For each $N' \in \mathfrak{I}'$ there is some $N \in \mathfrak{I}$ such that $g(N) \subset g(N')$. Equivalently, $N' \in \mathfrak{I}'$ implies $g^{-1}(N') \in \mathfrak{I}$.

Continuity of a function expresses the idea that points in the domain that are extremely close together should map into points that are not far apart. Informally, a *preference relation* is continuous if whenever x is strictly preferred to y then everything close to x is strictly preferred to everything close to y.

3. Continuous relation. The binary relation $R \subset X^2$ is continuous in topology $\mathfrak{I}$ if for each $x \in X$ the sets $\{ y \in X \,|\, (x,y) \in AR \}$ and $\{ y \in X \,|\, (y,x) \in AR \}$ are open.

The imposition of continuity can lead to meaningful results only if the topology employed is appropriate to the problem at hand. For instance, continuity is not at all restrictive in the discrete topology because every binary relation is continuous in that topology. The discrete topology makes a distinction between any two allocations of private goods that are arbitrarily close in Euclidean distance, but not all such distinctions can be made in practice. (See the discussion following Proposition 8 for a definition of the Euclidean topology.) If the set of alternatives from which a choice is to be made is finite then the discrete topology is appropriate. In fact, the discrete topology is implied in that case, as the next result confirms. As a consequence, continuity is not restrictive when a finite set of alternatives is being considered.

4. Proposition. If Y is a finite subset of a T_1-space then the relative topology on Y is discrete.

Proof. Suppose $Y = \{y_1, y_2, \ldots, y_m\}$. For each i and j such that $i \neq j$ there is an open set N_{ij} such that $y_i \in N_{ij}$ and $y_j \notin N_{ij}$. Therefore, $N_i \equiv Y \cap_{j \neq i} N_{ij} = \{y_i\}$. The set $\cap_{j \neq i} N_{ij}$ is open in the original space, by repeated application of 1b, and hence $\{y_i\}$ belongs to $\mathfrak{I}_Y$. Then $\mathfrak{I}_Y$ is discrete by property 1c. ∎

All social choice impossibility theorems assume that the set X of alternatives contains more than two members. This will be implicit in most of the theorems in this book by virtue of the following simple corollary of Proposition 4.

5. Proposition. A connected T_1-space is infinite unless it is a singleton.

Not only does the continuity assumption rule out pathological preference reversals, it also guarantees the existence of a maximal element if the space is compact and the relation is acyclic. By definition, a space is compact if every open cover has a finite subcover: An open cover is a subset Γ of $\mathfrak{I}$ such that $\cup \Gamma = X$. A finite subcover is a finite subset $\{N_1, N_2, \ldots, N_m\}$ of Γ such that $X = N_1 \cup N_2 \cup \ldots N_{m-1} \cup N_m$. In fact we can prove a much stronger result called the Bergstrom–Walker theorem. *Lower* continuity suffices for the existence of a maximal element. A binary relation R on X is lower continuous if $\{y \in X \mid (x,y) \in AR\}$ is open for each $x \in X$.

Note that the discrete topology is connected if and only if the space is a singleton.

6. Theorem. (Bergstrom, 1975; Walker, 1977.) If R is a lower continuous and acyclic binary relation on a compact space X then there exists some $x^* \in X$ such that $(y,x^*) \notin AR$ for any $y \in X$.

Proof. Suppose that there is no such maximal element x^*. Then the set $\Gamma = \{ Y_x \mid x \in X \}$ covers X if $Y_x = \{ y \in X \mid (x,y) \in AR \}$. Because X is compact there is a finite subcover, say Y_1, Y_2, ..., Y_m, where $Y_i = Y_x$ for $x = x_i$. Let y be R-maximal in $\{x_1, x_2, ..., x_m\}$. Such a maximal element exists by Proposition 1.18. But $\cup_{i=1}^{m} Y_i$ covers X so we must have $y \in Y_i$ for some i. Then $(x_i,y) \in AR$, contradicting the fact that y is R-maximal in $\{x_1, x_2, ..., x_m\}$. Therefore, the assumption that no maximal element exists must be abandoned. ∎

If R is a preorder the continuity requirement can be substantially weakened. See Campbell and Walker (1990), Tian (1990), or Tian and Zhou (1990). Most spaces of alternatives employed in economics are not compact but the relevant subspaces usually are. For example, budget sets are compact in the Euclidean topology if all prices are positive, and a set of feasible allocations for an economy is compact if the technology is realistic. (See the discussion following Proposition 8 below for a definition of the Euclidean topology.)

A topological space satisfies the *first axiom of countability* if for each $x \in X$ there is a countable family $\{N_1, N_2, ...\}$ of open sets such that each N_t contains x and each open set containing x also contains at least one member of the family. This means that $\{x\} = \cap_{t=1}^{\infty} N_t$ if the space is also T_1. The family $\{N_1, N_2, ...\}$ is called a neighbourhood base for x.

7. Proposition. If X is a T_1-space satisfying the first axiom of countability then each member of x can be expressed as the intersection of a countable family of open sets.

Proof. Choose arbitrary $x \in X$. First countability means that x has a countable neighbourhood base $\{N_1, N_2, ...\}$. Let $Z = \cap_{t=1}^{\infty} N_t$. If $y \neq x$ and the space is T_1 then there exists an open set U containing x but not y. Then $N_t \subset U$ for some t. Therefore, $y \notin Z$. Obviously, $x \in Z$. ∎

We will also need the following simple fact about first-countable spaces that are connected T_1-spaces.

8. Proposition: If X is a connected and first-countable T_1-space then for every $x \in X$ there is a sequence $\{x'\}$ in $X - \{x\}$ converging to x.

Proof. Let $\{N_1, N_2, ...\}$ be a neighbourhood base for x. Define

$\{U_1, U_2, \ldots\}$ by setting $U_t = \cap_{n=1}^{t} N_n$ for $t = 1, 2, \ldots$. Then $\{U_1, U_2, \ldots\}$ is also a neighbourhood base for x. If $U_t = \{x\}$ then U_t is closed because the space is T_1. But then U_t and $X - U_t$ are closed and disjoint subsets such that $U_t \cup (X - U_t) = X$ contradicting the connectedness of X. Therefore, $U_t \neq \{x\}$. Define $\{x'\}$ by choosing any $x' \in U_t - \{x\}$. Obviously, $\{x'\} \subset X - \{x\}$. If V is an open set containing x then there is some $U_n \subset V$. Because $U_t \subset U_n$ for all $t \geq n$ we have $x' \in U_n$ for all $t \geq n$. Therefore, the sequence $\{x'\}$ converges to x. ∎

If $(\Im_\lambda, X_\lambda)$, $\lambda \in \Lambda$, is an indexed family of topological spaces we can define a product topology on the Cartesian product set $X = \Pi_{\lambda \in \Lambda} X_\lambda$, which denotes the set of functions π on Λ such that $\pi_\lambda \equiv \pi(\lambda) \in X_\lambda$, $\forall \lambda \in \Lambda$. The product topology $\Im$ is obtained by first defining a *base* $\Re$ for the topology. $\Re$ is the family of subsets Y of X such that for each $\lambda \in \Lambda$ the set $\{\pi_\lambda \mid \pi \in Y\}$ belongs to $\Im_\lambda$ and is in fact equal to X_λ for all but a finite number of λ. Then $\Im$ is the family of sets obtained as unions of arbitrary subsets of $\Re$ plus the empty set.

The real line $\mathbf{E}$ with its usual topology generates the allocation spaces employed most often in economic theory. (A set N is open in $\mathbf{E}$ if and only if for each $x \in Y$ there is some strictly positive ϵ in $\mathbf{E}$ such that y belongs to N whenever $x - \epsilon < y < x + \epsilon$.) If k is a positive integer then $\mathbf{E}^k$ denotes k-dimensional Euclidean space, which is the Cartesian product set $\Pi_{\lambda \in \Lambda} X_\lambda$, for $\Lambda = \{1, 2, \ldots, k\}$ and $X_\lambda = \mathbf{E}$, endowed with the product topology. Let $\mathbf{E}^k_+$ denote the non-negative orthant of $\mathbf{E}^k$. That is $\mathbf{E}^k_+ = \{x \in \mathbf{E}^k \mid x \geq 0\}$. The strictly positive orthant is $\mathbf{E}^k_{++} = \{x \in \mathbf{E}^k \mid x \gg 0\}$, where $x \gg 0$ means that $x_c > 0$ for $c = 1, 2, \ldots, k$. Of course, x_c is the cth coordinate of x. Note that $\mathbf{E}^k_+$ is a topological space; the topology is the relative one derived from $\mathbf{E}^k$. A subset Y of $\mathbf{E}^k_+$ is open if and only if for each $x \in Y$ there is some $\epsilon > 0$ such that $y \in Y$ whenever $y \geq 0$ and $x_c - \epsilon < y_c < x_c + \epsilon$ for $c = 1, 2, \ldots, k$. Because $\mathbf{E}^k_{++}$ itself is open in $\mathbf{E}^k$ a subset of E^k_{++} is open in the relative topology if and only if it is open in the Euclidean topology on $\mathbf{E}^k$.

Now, let T denote the set of individuals in society. The allocation space is $[\mathbf{E}^k_+]^T$, denoted Ω, the space of allocations of the k private goods. Each x in Ω is a function from T into $\mathbf{E}^k_+$, and $x(t)$ denotes the commodity vector assigned to individual $t \in T$. Then $x_c(t)$ is the amount of commodity c allocated to individual t by x. Let Ω_0 denote $[\mathbf{E}^k_{++}]^T$, the set of allocations of strictly positive commodity vectors. A set is compact in $\mathbf{E}^k$ if and only if it is closed and bounded. (This is the Heine–Borel theorem: see Dugundji 1966: 233.)

A subset of a topological space X is called a G_δ set if it is the intersection of a countable collection of open sets. The space itself satisfies the first axiom of countability if $\{x\}$ is a G_δ set for each $x \in X$. When proving impossibility theorems for the domain of all profiles of continuous preorders on X we will need the fact that for distinct x and y in X we can find a real-valued continuous function g on X such that $g(x) < g(z) < g(y)$ for all $z \in X$ such that $x \neq z \neq y$. This will be possible if X is a normal space satisfying the first axiom of countability, as we now prove.

9. Theorem. (P. Urysohn.) If X is normal and Y and Z are disjoint closed subsets of X then there is a real-valued continuous function g on X such that $Y \subset g^{-1}(0)$, $Z \subset g^{-1}(1)$, and $0 \leq g(x) \leq 1$ $\forall$ $x \in X$.

Proof. Let K be the set of rationals of the form $k/2^m$, with k and m any non-negative integers satisfying $k/2^m \leq 1$. For each $\alpha \in K$ we will define an open subset $U(\alpha)$ of X such that the following two conditions hold for α, $\beta \in K$.

9a $\quad Y \subset U(\alpha), Z \cap U(\alpha) = \varnothing$.
9b $\quad \mathrm{clos}\, U(\alpha) \subset U(\beta)$ whenever $\alpha < \beta$.

Set $K_m = \{0/2^m, 1/2^m, 2/2^m, \ldots, 2^{m-1}/2^m, 2^m/2^m\}$. Let $U(1)$ be the complement of Z. Because X is normal there exist open sets $U(0)$ and N such that $Y \subset U(0)$, $Z \subset N$, and $U(0) \cap N = \varnothing$. Then $U(0)$ is a subset of $X - N$, the complement of N. By definition, $X - N$ is a closed set, so $\mathrm{clos}\, U(0) \subset \mathrm{clos}\,(X - N) = X - N$. Then $\mathrm{clos}\, U(0) \subset X - Z$ because $X - N \subset X - Z$. To summarize, we have $Y \subset U(0) \subset \mathrm{clos}\, U(0) \subset X - Z = U(1)$. K_0 satisfies 9a and 9b. We will define $U(\alpha)$ inductively. If k is even we have $\alpha = (k/2)/2^{m-1} \in K_{m-1}$ so we need only define $U(\alpha)$ for k odd, assuming that $U(\alpha)$ has been defined for all $\alpha \in K_{m-1}$ in accordance with 9a and 9b. Then for k odd we have $\mathrm{clos}\, U((k-1)/2^m) \subset U((k+1)/2^m)$ by 9b. By normality, there exist disjoint open sets N and M satisfying $\mathrm{clos}\, U((k-1)/2^m) \subset N$ and $[X - U((k+1)/2^m)] \subset M$. Therefore

$$\mathrm{clos}\, U((k-1)/2^m) \subset N \subset \mathrm{clos}\, N \subset U((k+1)/2^m).$$

Set $U(k/2^m) = N$. Then K_m satisfies 9a and 9b, and $K = \cup_{m=1}^{\infty} K_m$.

Now define g by setting $U(1) = X$ instead of $X - Z$ and

$$g(x) = \inf\{\alpha \in K \mid x \in U(\alpha)\}.$$

We have $0 \le g(x) \le 1$ and $g(Z) = \{1\}$ because $x \in Z \cap U(\alpha)$ implies $\alpha = 1$. Also, $g(Y) = \{0\}$ because $Y \subset U(\alpha)$ for all $\alpha \in K$. To show that g is continuous choose $x \in X$ and let ϵ be an arbitrary positive number. We need to find an open set N in X such that $x \in N$ and $g(x) - \epsilon < g(y) < g(x) + \epsilon$ for all $y \in N$. If $g(x) = 0$ choose $\beta \in K$ such that $g(x) < \beta < g(x) + \epsilon$. By definition, $g(y) \le \beta$ for all $y \in U(\beta)$ and we can set $N = U(\beta)$. If $g(x) = 1$ choose $\alpha \in K$ such that $g(x) - \epsilon < \alpha < g(x)$. If $y \notin \text{clos} U(\alpha)$ we have $g(y) \ge \alpha$ because 9b implies $U(\lambda) \subset U(\alpha)$ for all $\lambda < \alpha$; we can let N be the complement of $\text{clos} U(\alpha)$. If $0 < g(x) < 1$ choose α and β so that $g(x) - \epsilon < \alpha < g(x) < \beta < g(x) + \epsilon$ and let N be the set of points in $U(\beta)$ that do not belong to the closure of $U(\alpha)$. Then $\alpha \le g(y) \le \beta$ for all y in N. $X - \text{clos} U(\alpha)$ is open and thus so is $U(\beta) \cap [X - \text{clos} U(\alpha)] = N$. We have $x \in U(\beta)$ because $g(x) < \beta$, and for $\alpha < \delta < g(x)$ we must have $x \notin U(\delta)$ by definition of g. Therefore, $x \notin \text{clos} U(\alpha)$ by 9b. Therefore, N is an open set containing x. This establishes the continuity of g. ∎

A set is a G_δ set if it can be expressed as the intersection of a countable collection of open sets. If Y and Z are closed G_δ sets we can actually find a continuous g such that $g^{-1}(0) = Y$, $g^{-1}(1) = Z$, and $0 \le g(x) \le 1$ for all x in X. First we prove an intermediate result.

10. Corollary. If Y is a G_δ set and Y and Z are closed disjoint subsets of a normal topological space X then there exists a continuous real-valued function g on X such that $g^{-1}(0) = Y$, $g(Z) = \{1\}$, and $0 \le g(x) \le 1$ for all $x \in X$.

Proof. Let $Y = \cap_{i=1}^{\infty} N_i$ with each N_i open in X. Because X is normal we can assume that $Z \cap N_1 = \varnothing$. If we let $M_i = N_1 \cap N_2 \cap \ldots \cap N_i$ then $Y = \cap_{i=1}^{\infty} M_i$ and $M_i \subset M_j$ if $i > j$. Each M_i is open as the finite intersection of open sets. Let g_i be a continuous real-valued function on X such that $g_i(Y) = \{0\}$, $g_i(X - M_i) = \{1\}$ and $0 \le g_i(x) \le 1 \; \forall x \in X$ (Theorem 9). Define g by setting $g(x) = \Sigma_{i=1}^{\infty} 2^{-i} g_i(x)$. Clearly $g(x) = 0$ for all $x \in Y$. And $g(x) = 1$ for all $x \in Z$ because $M_i \subset M_1$ and $Z \subset X - M_1$.

Suppose $x \in X$ but $x \notin Y$. Because $Y = \cap_{i=1}^{\infty} M_i$ we have $x \notin M_i$ for some i and hence $g_i(x) = 1$ which implies $g(x) > 0$. The proof that g is continuous is postponed until Proposition 12 below. ■

11. Corollary. If Y and Z are closed and disjoint G_δ subsets of a normal topological space X then there exists a continuous real-valued function g such that $g^{-1}(0) = Y$, $g^{-1}(1) = Z$, and $0 \le g(x) \le 1 \, \forall x \in X$.

Proof. By Corollary 10 there exist continuous real-valued functions α and β such that $\alpha^{-1}(0) = \{Y\}$, $\beta^{-1}(0) = \{Z\}$ and $0 \le \alpha(x) \le 1$, $0 \le \beta(x) \le 1 \, \forall x \in X$. Let $g(x) = \alpha(x)/[\alpha(x) + \beta(x)]$ define g. The denominator is never zero because Y and Z are disjoint. Therefore, g is continuous. Clearly, $g(x) = 0$ if and only if $x \in Y$. Obviously, $g(x) = 1$ for all $x \in Z$. If $x \notin Z$ then $\beta(x) > 0$ and hence $g(x) < 1$. ■

12. Proposition. Let g_i $(i = 1,2, \ldots)$ be an infinite sequence of real-valued functions on the topological space X. If α_i is a real number $(i = 1,2, \ldots)$, $\Sigma_{i=1}^{\infty} \alpha_i$ converges, and $|g_i(x)| \le \alpha_i$ for all $x \in X$ and $i = 1,2, \ldots$ then the function defined by $g(x) = \Sigma_{i=1}^{\infty} g_i(x)$ exists and is continuous.

Proof. Set $\sigma_i(x) = g_1(x) + g_2(x) + \ldots + g_i(x)$ and $\beta = \Sigma_{i=1}^{\infty} \alpha_i$. Let ϵ be an arbitrary positive number. Then there is an integer j such that $|\alpha_1 + \alpha_2 + \ldots + \alpha_i - \beta| < \epsilon/2$ for all $i \ge j$. That is

12a $-\epsilon/2 < \alpha_1 + \ldots + \alpha_i - \beta < \epsilon/2$ and
12b $-\epsilon/2 < \beta - \alpha_1 - \ldots - \alpha_j < \epsilon/2$

if $i \ge j$. Set $\delta_i = \alpha_{j+1} + \alpha_{j+2} + \ldots + \alpha_i$. Then 12a and 12b imply $-\epsilon < \delta_i < \epsilon$ for $i > j$. Because $-\alpha_i \le g_i(x) \le \alpha_i$ for all i we have $\sigma_i(x) - \sigma_j(x) \le \delta_i < \epsilon$ and $\sigma_i(x) - \sigma_j(x) \ge -\delta_i > -\epsilon$ for all $i > j$. Therefore, $|\sigma_i(x) - \sigma_j(x)| < \epsilon$ for all $i \ge j$ and all $x \in X$. This means that σ_i converges uniformly to g. In particular, g is well defined. For any $x \in X$

$$|g(y) - g(x)| \le |g(y) - \sigma_i(y)| + |\sigma_i(y) - \sigma_i(x)| + |\sigma_i(x) - g(x)|$$

for all $y \in X$ and all i. Let ϵ be an arbitrary positive number. Because σ_i converges uniformly to g there exists an integer j such that $i \ge j$ implies $|g(y) - \sigma_i(y)| < \epsilon/3$ and $|\sigma_i(x) - g(x)| < \epsilon/3$. Any *finite* sum of continuous functions is continuous so there exists a neighbourhood $N(x)$ of x such that $|\sigma_j(y) - \sigma_j(x)| < \epsilon/3$ for all $y \in N(x)$. Therefore, $|g(y) - g(x)| < \epsilon$ for all $y \in N(x)$ and hence g is continuous. ∎

Recall that the topological space $(X, \mathfrak{I})$ is connected if one cannot find open sets Z_1 and Z_2 such that $X = Z_1 \cup Z_2$, $Z_1 \cap Z_2 = \emptyset$, and $Z_1 \neq \emptyset \neq Z_2$. Simply put, a connected space cannot be expressed as the union of non-empty and disjoint open sets. If $(X, \mathfrak{I})$ is any topological space we say that the subset Y of X is a connected *set* if $(Y, \mathfrak{I}_Y)$ is a connected space. ($\mathfrak{I}_Y$ is the relative topology.) Therefore, the set Y is connected if and only if there do not exist open sets Z_1 and Z_2 in $\mathfrak{I}$ such that Y is a subset of $Z_1 \cup Z_2$, and $Z_1 \cap Z_2 \cap Y = \emptyset$, and $Z_1 \cap Y \neq \emptyset \neq Z_2 \cap Y$. We need to prove that certain subsets of Ω are connected. These subsets play a central role in Part IV in extending the impossibility theorems to economic environments.

We begin by demonstrating that any interval in $\mathbf{E}$ is a connected set. An interval is a set of the form $(\alpha, \beta) = \{\lambda \in \mathbf{E} | \alpha < \lambda < \beta\}$, or $[\alpha, \beta] = \{\lambda \in \mathbf{E} | \alpha \leq \lambda \leq \beta\}$, or $(\alpha, \beta] = \{\lambda \in \mathbf{E} | \alpha < \lambda \leq \beta\}$, or $[\alpha, \beta) = \{\lambda \in \mathbf{E} | \alpha \leq \lambda < \beta\}$. Of course α and β must satisfy $\alpha \leq \beta$. We allow $\alpha = \beta$ only in the case of the closed interval $[\alpha, \beta]$ in which case we have a degenerate interval consisting of the single point α. And α and β must be real numbers unless $\alpha = -\infty$ or $\beta = +\infty$.

13. Proposition. An interval in $\mathbf{E}$ is connected.

Proof. Let I be an interval. Suppose that I is not connected and $I = Y \cup Z$ with $Y \cap Z = \emptyset$, $Y \neq \emptyset \neq Z$, and Y and Z open in the relative topology on I. Choose $\alpha \in Y$ and $\beta \in Z$. Without loss of generality assume that $\alpha < \beta$. I is an interval so $(\alpha, \beta) \subset I$. Because Y is open in the relative topology on I there is some $\lambda > \alpha$ such that $(\alpha, \lambda) \subset Y$. Set $\delta = \sup\{\lambda \in \mathbf{E} | (\alpha, \lambda) \subset Y\}$. Then $\alpha < \delta \leq \beta$. If $\delta \in Z$ then for all $\epsilon > 0$ we have, by definition of δ, $(\delta - \epsilon, \delta + \epsilon) \cap Y \neq \emptyset$, contradicting openness of Z in I and the fact that $Y \cap Z = \emptyset$. Therefore, $\delta \in Y$ and thus $\delta < \beta$. Because Y is open in I there is some $\epsilon > 0$ such that $(\delta - \epsilon, \delta + \epsilon) \cap I \subset Y$. Let $\epsilon' = \min\{\epsilon, (\delta + \alpha)/2\}$. Then $(\alpha, \delta + \epsilon') \subset I$ so $(\alpha, \delta + \epsilon') \subset Y$. Because $\epsilon' > 0$ this contradicts the definition of δ. ∎

If $g: X \to X'$ is a function the *image* of g is the set $g(X) = \{g(x) | x \in X\}$.

14. Proposition. The continuous image of a connected set is connected.

Proof. Let $(X,\mathfrak{I})$ and $(X',\mathfrak{I}')$ be topological spaces. Suppose that $g: X \to X'$ is a continuous function but $g(X)$ is not connected. Then there exist $\mathfrak{I}'$-open subsets Y' and Z' of X' such that $Y' \cap Z' \cap g(X) = \varnothing$, $g(X) \subset [Y' \cup Z']$ and $Y' \cap g(X) \neq \varnothing \neq Z' \cap g(X)$. Because g is continuous the sets $Y = g^{-1}(Y')$ and $Z = g^{-1}(Z')$ are open in $\mathfrak{I}$. Because $g(X') \subset [Y' \cup Z']$ we have $Y \cup Z = X$. Because $Y' \cap g(X) \neq \varnothing \neq Z' \cap g(X)$ we have $Y \neq \varnothing \neq Z$. Because $Y' \cap Z' \cap g(X) = \varnothing$ we have $Y \cap Z = \varnothing$. But we have contradicted the connectedness of $(X,\mathfrak{I})$. Therefore $g(X)$ must be connected. ∎

15. Proposition. For any $\alpha > 0$ and β in $\mathbf{E}$ the set $C = \{x \in \mathbf{E}^2 | x \geqslant 0$ and $\log x_1 + \alpha \log x_2 = \beta\}$ is connected.

Proof. According to Propositions 13 and 14 we need only find an interval I in $\mathbf{E}$ and a continuous function $g:I \to \mathbf{E}^2$ such that $g(I) = C$. Simply take $I = (0,\infty) = \{\lambda \in \mathbf{E} | 0 < \lambda\}$ and set $g(\lambda) = (\lambda, (\delta/\lambda)^{1/\alpha})$ for δ satisfying $\beta = \log\delta$. If $x = g(\lambda)$ we have $x_1 x_2^\alpha = \delta$ and hence $\log x_1 + \alpha \log x_2 = \beta$. Therefore, $g(I) = C$. The function g is obviously continuous, and I is connected by Proposition 13. Therefore, C is connected by Proposition 14. ∎

We need to know that the product of subspaces such as C is also connected. (Such products will be called Cobb–Douglas spaces.) Two new lemmas are required for this purpose; they will also be put to other uses.

16. Lemma. Let $(X,\mathfrak{I})$ be any topological space. If Y is the union of a family of connected sets in X having one point in common then Y is also connected.

Proof. Suppose Z_λ is a connected subset of X for each $\lambda \in \Lambda$, and $\cap_{\lambda \in \Lambda} Z_\lambda$ contains at least one point, say z. Set $Y = \cup_{\lambda \in \Lambda} Z_\lambda$. If Y is not connected there are open sets V_1 and V_2 in $\mathfrak{I}$ such that $Y \subset [V_1 \cup V_2]$, $V_1 \cap Y \neq \varnothing \neq V_2 \cap Y$, and $V_1 \cap V_2 \cap Y = \varnothing$. Suppose $z \in V_1$. If $V_2 \cap Z_\lambda \neq \varnothing$ we have a contradiction: $Z_\lambda \subset [V_1 \cup V_2]$, $V_1 \cap Z_\lambda \neq \varnothing \neq V_2 \cap Z_\lambda$, and $V_1 \cap V_2 \cap Z_\lambda = \varnothing$. Because Z_λ is connected and $\lambda \in \Lambda$ was chosen arbitrarily, we must have $Z_\lambda \subset V_1$ for all $\lambda \in \Lambda$. This contradicts $V_2 \cap Y \neq \varnothing$ and $V_1 \cap V_2 \cap Y = \varnothing$. Therefore, Y must be connected. ∎

17. Lemma. Let Y be a subset of the topological space $(X, \mathfrak{I})$. If Z is a connected subset of X and $Z \subset Y \subset \text{clos} Z$ then Y is also connected.

Proof. Suppose that V_1 and V_2 are non-empty open sets in X, $V_1 \cap V_2 \cap Y = \varnothing$ and Y is a subset of $V_1 \cup V_2$. Then $V_1 \cap V_2 \cap Z = \varnothing$ and Z is a subset of $V_1 \cup V_2$. Because Z is connected we must have $Z \subset V_i$ for some i. Suppose $y \in V_j \cap Y$ for $j \neq i$. Then y belongs to the closure of Z. (By hypothesis $Y \subset \text{clos} Z$.) But V_j is open so we must have $V_j \cap Z \neq \varnothing$ contradicting $Z \subset V_i$ and $V_i \cap V_2 \cap Z = \varnothing$. Therefore, $Y \subset V_i$ and Y must be connected. ∎

18. Proposition. Let $(X_\lambda, \mathfrak{I}_\lambda)$ be a connected topological space for each λ in the index set Λ. Then the product set $X = \Pi_{\lambda \in \Lambda} X_\lambda$ is connected in the product topology $\mathfrak{I}$ on X.

Proof. Choose $x^* \in X$. We will prove the following statement by induction.

18a If $x \in X$ and $x(\lambda) \neq x^*(\lambda)$ for at most a finite number n of coordinates λ then there is a connected set $Y \subset X$ containing both x^* and x.

Consider the case $n = 1$. Suppose $\alpha \in \Lambda$, $x \in X$, and $x(\lambda) = x^*(\lambda)$ for all $\lambda \neq \alpha$. Let $Y = \{y \in X | y(\lambda) = x^*(\lambda) \, \forall \lambda \neq \alpha\}$. Then Y is connected because X_α is connected and $Y = X_\alpha \times \Pi_{\lambda \neq \alpha} \{x^*(\lambda)\}$. Y is the image of the continuous function $g: X_\alpha \to Y$ on the connected set X_α, where g is defined by

$$g(x(\alpha)) = y \text{ for } y(\alpha) = x(\alpha) \text{ and } y(\lambda) = x^*(\lambda), \, \forall \lambda \neq \alpha.$$

Then Y is connected by Proposition 14.

Now, suppose that 18a holds for $n = 1, 2, \ldots, m$. Choose $x \in X$ such that $x(\lambda) \neq x^*(\lambda)$ for $m + 1$ coordinates λ. Choose $y \in X$ so that $y(\lambda) = x^*(\lambda)$ for exactly one λ such that $x(\lambda) \neq x^*(\lambda)$ and $y(\lambda) = x(\lambda)$ for all other $\lambda \in \Lambda$. Then there is a connected set Y containing x and y. But y and x^* differ in only m coordinates and by the induction hypothesis there is some connected set Z containing y and x^*. Then $y \in Y \cap Z$ and hence $Y \cup Z$ is connected by Lemma 16; we have proved 18a for $n = m + 1$.

Now, let Y^* be the union of all connected subsets of X containing x^*. By Lemma 16 the set Y^* is connected. By 18a the set Y^* contains $Z^* = \{x \in X | x(\lambda) \neq x^*(\lambda) \text{ for a finite number of } \lambda \text{ at most}\}$. Suppose that $\text{clos} Z^* = X$. We have $\text{clos} Z^* \subset \text{clos} Y^*$ because $Z^* \subset Y^*$.

Then $\mathrm{clos}Z^* = X$ implies $\mathrm{clos}Y^* = X$. The closure of a connected set is connected by Lemma 17. Therefore, $\mathrm{clos}Y^*$ is connected and thus $\mathrm{clos}Z^* = X$ implies that X is connected. It remains to prove that the closure of Z^* actually equals X.

Suppose $x \in X$ and $N(x)$ is a neighbourhood of x. By definition of the product topology we must have $\{x(\lambda)\,|\,x\in N(x)\,\} = X_\lambda$ for all but a finite number of $\lambda \in \Lambda$. Therefore, there is some $y \in N(x)$ such that $y(x) = x^*$ for all but a finite number of λ. Therefore, $y \in Z^*$. Every neighbourhood of x contains a member of Z^*. Therefore, $\mathrm{clos}Z^* = X$. ∎

We now have all but one of the results that are required for the social choice theorems to follow. We conclude this chapter with a remark on continuity and indifference.

Many theorems on social choice assume that individual preferences belong to the family of linear orderings (1.24) on the outcome set X. (For example, Kalai and Muller 1977, and Kalai and Ritz 1980.) This assumption is inconsistent with *continuity* of a preorder on $\mathbf{E}_+^k$ if $k > 1$. We now demonstrate this fact. The *null relation* $X^2 = \{\,(x,y)\,|\,x\in X \text{ and } y\in X\,\}$, which sets everything indifferent to everything else, is a linear order if and only if X is a singleton. Consider then the case of a continuous preorder R on a connected space X such that $(x,y) \in AR$ for some choice of x and y. First, we prove that there is some alternative z such that (x,z) and (z,y) both belong to AR.

19. Lemma. If $(x,y) \in AR$ and R is a continuous preorder on a connected topological space X then there is some $z \in X$ such that $(x,z) \in AR$ and $(z,y) \in AR$.

Proof. Set $Y_1 = \{z\in X\,|\,(z,y)\in AR\,\}$ and $Y_2 = \{z\in X\,|\,(x,z)\in AR\,\}$. If $z \notin Y_1$ then $(y,z) \in R$ and hence $(x,z) \in AR$ because $(x,y)\in AR$ and R is a preorder (1.27). Therefore, $z \notin Y_1$ implies $z \in Y_2$ and hence $X = Y_1 \cup Y_2$. We have $x \in Y_1$ and $y \in Y_2$ and thus neither set is empty. They are both open by definition of continuity of a preorder. Because X is connected we must have $Y_1 \cap Y_2 \neq \varnothing$. But (x,z) and (z,y) both belong to AR for any z in $Y_1 \cap Y_2$. ∎

Now, suppose R is a non-null and continuous preorder on $X = \mathbf{E}_+^k$, which is a connected topological space by Propositions 13 and 18. Assume that $k > 1$. Because R is non-null we have $(x,y) \in AR$ for some choice of x and y in X. Therefore, there is some $z \in X$ such that (x,z) and (z,y) both belong to AR by Lemma 19.

(Applying this lemma repeatedly we can find an infinite list of alternatives, none of which is indifferent to any other on the list.) For each choice of $v \in X$ distinct from x, y, and z we have a connected curve C defined as the union of the line segment connecting x and v with the line segment connecting v and y (Propositions 13 and 14 and Lemma 16). We can construct an infinite number of curves in this fashion, and in such a way that any two have only the end-points x and y in common and none of the curves contains z. Set $Y_1 = \{ b \in C \,|\, (b,z) \in AR \}$ and $Y_2 = \{ b \in C \,|\, (z,b) \in AR \}$. Clearly, $Y_1 \cap Y_2 = \varnothing$ and $Y_1 \neq \varnothing \neq Y_2$. Because R is continuous the set $Z_1 = \{ b \in X \,|\, (b,z) \in AR \}$ is open in the topology on X, and so is $Z_2 = \{ b \in X \,|\, (z,b) \in AR \}$. Therefore, $Y_i = Z_i \cap C$ is open in the relative topology on C. Because C is connected we cannot have $C = Y_1 \cup Y_2$. Choose any b in C that does not belong to Y_1 or Y_2. We have $(b,z) \in SR$ by Proposition 1.9. Each distinct C will contain a different member of z's equivalence class, SRz (1.6 and 1.30). The only property of $\mathbf{E}_+^k$ that we used that would not be shared by every connected space is the existence of a connected curve containing x and y but not z. For example, this would not be possible on the real line with R representing the usual ordering: If $x > z > y$ then the line connecting x and y will pass through z. But if the connected space admits a connected curve joining two points and not containing the third point then a continuous preorder on that space cannot be linear.

3 Social Welfare Functions

THIS chapter introduces the basic apparatus and terminology of social choice theory as well as the criteria by which collective choice will be judged. Begin with X, a universal space of outcomes. That is, X is a set of alternatives endowed with some topology, and all of the sets of interest will be subsets of X endowed with the relative topology. Given $Y \subset X$ let $B(Y)$ denote the family of complete and continuous binary relations on Y. Specifically, R belongs to $B(Y)$ if and only if $R \subset Y^2$ and for all $x,y \in Y$, $(x,y) \notin R$ implies $(y,x) \in R$, and the sets $\{z \in Y \mid (z,x) \in AR\}$ and $\{z \in Y \mid (x,z) AR\}$ are open in the relative topology on Y. Recall that $AR = \{(x,y) \in R \mid (y,x) \notin R\}$, the asymmetric factor of R. By Proposition 2.4, if Y is finite and X is a T_1-space then Y has the discrete topology. Because every set is open in the discrete topology, $B(Y)$ is just the set of complete binary relations on Y in that case.

Let $B^*(Y)$ be the family of complete, continuous, and acyclic binary relations on Y. That is, R belongs to $B^*(Y)$ if and only if $R \in B(Y)$ and for every finite subset $\{y^1, y^2, \ldots, y^m\}$ of Y, $(y^i, y^{i+1}) \in AR$ for $i = 1, 2, \ldots, m-1$ implies $y^1 \neq y^m$. $Q(Y)$ denotes the family of complete, continuous, and quasitransitive binary relations on Y. $Q(Y) = \{R \in B(Y) \mid AR$ is transitive$\}$. Finally, $P(Y)$ is the set of continuous preorders on Y. $P(Y) = \{R \in B(Y) \mid R$ is transitive$\}$. Recall that $P(Y) \subset Q(Y) \subset B^*(Y)$. [Propositions 1.15, 1.16, and 1.18.]

Let T denote the set of individuals in society. T will usually be a finite set, but infinite societies will receive some consideration. A *domain* for a social choice problem involving the alternative set X is some subset D of $P(X)^T$. *Individual* preference will always be assumed to be complete and transitive. Each member of $P(X)^T$ is, by definition, a function $p: T \rightarrow P(X)$, called a *profile*: The profile p assigns the continuous preorder $p(t)$ to individual $t \in T$, and $p(t)$ is interpreted as t's preference scheme in situation p. A domain, then, is some set of profiles. The only domains that will be considered in this book are product sets. That is, for each $t \in T$ there is some $D_t \subset P(X)$ such that $D = \Pi_{t \in T} D_t$. In other words, p belongs to D if and only if $p(t) \in D_t$ for each $t \in T$. If D is defined for X, and Y is a subspace of X we define $D(Y)$, the restriction of D to Y, in the natural way:

$$D(Y) = \{p \in P(Y)^T \mid \text{there is some } q \in D \text{ such that}$$
$$p(t) = q(t) \cap Y^2 \text{ for all } t \in T\}.$$

Note that $R \cap Y^2$ is continuous in the relative topology on Y if R belongs to $B(X)$ and Y is a subset of X.

A *social welfare function f* for outcome space X and domain D is a function from D into $B(X)$. For each $p \in D, f(p)$ is a complete and continuous binary relation on X, interpreted as the social preference scheme determined by f when individual preferences are specified by p. We want to determine a social preference scheme by employing democratic values to incorporate information about individual preferences in a way that has general community acceptance. Because we cannot immediately see how that should be done, we begin slowly by bringing some elementary normative criteria to bear on the social welfare function f. (For example, f cannot depend exclusively on person 1's preferences.)

The first axiom is a standard efficiency condition which is a limited expression of the idea that individual wants should govern social choice. Everyone has some vague notion of what it means for social choice to be governed by individual wants, but it is not at all clear that anyone would be capable of formalizing the notion. Even if that were possible, the various individual definitions would be sure to conflict to some extent. However, it is possible to express a partial responsiveness condition with precision and with some confidence that the condition would be unanimously endorsed. If everyone agrees that x is superior to y then x should rank above y in the social preference relation. This does not immediately give rise to a particular social choice procedure, due to the fact that a change in the status quo will usually harm some individuals and benefit others. On the other hand, in cases when there *is* unanimity, society should respond accordingly.

1. The Pareto criterion. For all $x,y \in X$, if $(x,y) \in Ap(t)$ for all $t \in T$ then $(x,y) \in Af(p)$.

In words, whatever the profile and whatever the pair of alternatives being presented for consideration, if there is one alternative that is strictly preferred to another by absolutely everyone then the former must rank strictly above the latter in the social preference scheme. In 1896 the Italian economist Vilfredo Pareto proposed the far-reaching criterion that now bears his name. It rules out every conceivable kind of waste. In fact, it is the perfect definition of waste. There is waste somewhere in the system if y is the outcome and absolutely everyone prefers the feasible outcome x to y (Pareto 1896). Although the criterion was first employed in 1881 by F.Y. Edgeworth, he only applied it to an exchange economy. Pareto clearly had an understanding of the scope of efficiency; in 1909 he demonstrated that a competitive equilibrium is Pareto optimal (Pareto 1909). See Kirman (1987) for an excellent assessment of Pareto's contribution.

Even though the Pareto criterion does not appear to be very demanding it will be found to be in conflict with other modest requirements and we are forced to consider weak versions of the Pareto criterion. The first of these is *strict non-imposition*, which requires, for each distinct pair of alternatives x and y that are not restricted in some way by the domain D, the existence of at least one profile p in the domain such that x ranks strictly above y in the social preference scheme $f(p)$ and the existence of at least one profile p' in the domain such that y ranks strictly above x in the social preference scheme $f(p')$. The domain qualification is essential because we could have y ranking at the top of individual preference scheme $p(t)$ for each $t \in T$ and each $p \in D$. Then it would be unreasonable to insist that each $x \neq y$ ranks strictly above y in at least one situation. In general, we say that D is a priori *unrestricted* with respect to x and y, or that $\{x, y\}$ is *free*, if $D(\{x,y\}) = P(\{x,y\})^{\mathrm{T}}$ and $x \neq y$.

2. Strict non-imposition. For all x, $y \in X$ such that $x \neq y$ and $D(\{x,y\}) = P(\{x,y\})^{\mathrm{T}}$ there exists some $p \in D$ such that $(x,y) \in Af(p)$.

Strict non-imposition is implied by the requirement that x rank strictly above y in the social preference scheme whenever x gives everyone substantially more benefit than y, but it is formulated in a way that avoids the necessity of determining what it means for someone to receive substantially more benefit from x than from y. If we merely insist upon x ranking at least as high as y in some situation we have the following slightly weaker requirement.

3. Non-imposition. For all $x,y \in X$ such that $D(\{x,y\}) = P(\{x,y\})^\mathrm{T}$ there exists some $p \in D$ such that $(x,y) \in f(p)$.

An alternative weak version of strict non-imposition is *minimal responsiveness*: For each alternative x there exist $y \in X$ and two profiles p and p' such that $(x,y) \in Af(p)$ and $(y,x) \in Af(p')$, unless x and y are a priori restricted for all $y \neq x$. That is, for each x there is some y that ranks below x in the social preference scheme in some situation, and above x in another situation, unless x and y are a priori restricted for all $y \neq x$.

4. Minimal responsiveness. For all $x \in X$ there exist $y \in X - \{x\}$ and $p, p' \in D$ such that $(x,y) \in Af(p)$ and $(y,x) \in Af(p')$ unless $D(\{x,y\}) \neq P(\{x,y\})^\mathrm{T}$ for all $y \neq x$.

The weakest efficiency condition of all simply rules out constant social welfare functions. The social welfare function $f\colon D \to B(X)$ is *constant* if $f(p) = f(p')$ for all $p, p' \in D$. In plain words, social preference is completely unresponsive to individual preference. Requiring f to be non-constant is the weakest conceivable responsiveness criterion. As we will see, there are uncontrived social choice problems for which even this condition is in conflict with the most elementary notion of equity. In particular, if Ω is the space of allocations of private goods and W is the associated domain of classical economic preferences then a non-dictatorial social welfare function $f\colon W \to P(\Omega)$ will fail a very weak test of incentive compatibility unless it is constant, and hence trivially incentive-compatible. In other words, the only incentive-compatible functions from W into $P(\Omega)$ are either constant or dictatorial. (Weak incentive compatibility is defined in Chapter 13. It concerns the existence of a game form or mechanism that can be used to generate, at equilibrium points, all members of a feasible set that are top-ranked in terms of social preference.)

All of the above efficiency conditions are implied by the Pareto criterion. If $D(\{x,y\}) = P(\{x,y\})^\mathrm{T}$ and $x \neq y$ then there exists a profile $p \in D$ such that $(x,y) \in \bigcap_{t \in T} Ap(t)$ and thus $(x,y) \in Af(p)$ by the Pareto criterion. Therefore, the Pareto criterion implies strict non-imposition and minimal responsiveness. This implicitly assumes that $\{x\}$ and $\{y\}$ are open in the relative topology on $\{x,y\}$ to ensure that $(x,y) \in Ap(t)$ is consistent with continuity of $p(t) \cap \{x,y\}^2$, but this will be the case with any T_1-space X, and hence with any space considered in this book. The Pareto criterion implies that f is non-constant if there is even one pair x,y in X such that $(x,y) \in \cap_{t \in T} Ap(t)$ for some $p \in D$ and $(y,x) \in \cap_{t \in T} Ap'(t)$ for some $p' \in D$. This simple condition will be satisfied by all of the domains considered in the following pages.

Now we come to an axiom that has received much more criticism than that Pareto criterion: *independence of irrelevant alternatives*. This axiom made its first appearance in Arrow (1951), and will be referred to throughout as *Arrow's independence axiom*. It requires the social ordering of x and y to be the same in two situations if, person by person, the individual orderings are the same in those two situations.

5. Arrow's independence axiom. For all $p,p' \in D$ and all $x, y \in X$, if $p(t) \cap \{x,y\}^2 = p'(t) \cap \{x,y\}^2$ for all $t \in T$ then $f(p) \cap \{x,y\}^2 = f(p') \cap \{x,y\}^2$.

This axiom allows us to define the social welfare function $f|Y$, the restriction of f to an arbitrary subset Y of X. (See 8.16 for a formal definition of $f|Y$.) Arrow's independence axiom will be in force at every turn with the exception of the Chapters 7 and 13 dealing with incentive compatibility. In Chapter 13 we will show that Arrow's independence axiom is actually implied by incentive compatibility. This gives a strong justification for imposing the independence axiom at an early stage. (Not only is the ethical appropriateness of this axiom controversial, but there is disagreement about Arrow's original definition and its justification. The paper by Bordes and Tideman (1991) goes a long way towards clarifying the issue.)

The next series of properties are ones that we want a social welfare function to avoid, but before introducing them we present some new terminology.

A *coalition* is a non-empty subset of T. Let $f: D \to B(X)$ be a social welfare function. All of the following definitions refer to f.

6. Decisiveness. Coalition $I \subset T$ is decisive for the ordered pair (x,y) if we have $(x,y) \in Af(p)$ whenever $p \in D$ and $(x,y) \in Ap(t)$ for all $t \in I$. Coalition I is globally decisive (or just decisive) if it is decisive for all ordered pairs.

In words, I is decisive if x ranks above y in the social preference scheme whenever all members of I strictly prefer x to y. Formally, I is decisive if $\cap_{t \in I} Ap(t) \subset Af(p)$ for all $p \in D$. Suppose, for example, that person 1 is a *complete dictator*: that is, for all $p \in D$

we have $f(p) = p(1)$. The social preference scheme is always identical to person 1's preference ordering. Then any coalition containing individual 1 is decisive over all pairs. Clearly, more information is conveyed by the statement 'coalition $\{1\}$ is decisive' than by the statement 'coalition T is decisive'. Typically, the smallest decisive coalition will have power that larger decisive coalitions do not possess. This additional power is *veto power*.

7. Veto power. Individual t has veto power if $Ap(t) \subset f(p) \, \forall p \in D$.

This means that x cannot rank above y in the social preference scheme if the individual possessing veto power strictly prefers y to x. Obviously, a complete dictator has veto power but a dictator will have veto power even if he is not a complete dictator.

8. Dictatorship. Individual t is a dictator, and the social welfare function is dictatorial if for all $x,y \in X$ and $p \in D$ we have $(x,y) \in Af(p)$ whenever $(x,y) \in Ap(t)$. That is, a dictator is an individual who is globally decisive. Individual t is a complete dictator if $f(p) = p(t)$ for all $t \in T$, in which case f is said to be completely dictatorial.

Consider the following example.

9. Example. $T = \{1,2\}$ and X is a discrete space with at least three alternatives. Set $D = P(X)^T$ and $(x,y) \in Af(p)$ if and only if $(x,y) \in Ap(1)$, or $(x,y) \in Sp(1) \cap Ap(2)$.

In words, x ranks above y socially if 1 strictly prefers x to y, or 1 is indifferent and individual 2 strictly prefers x to y. Individual 1 is a dictator according to Definition 8 and thus has veto power. But 1 is not a complete dictator because there are situations in which x ranks above y socially although 1 is indifferent between x and y. Notice that the social welfare function defined by Example 9 maps $P(X)^T$ into $P(X)$. If (x,y) and (y,z) belong to $f(p)$ then (x,y) and (y,z) both belong to $p(1)$, by definition. Then $(x,z) \in Ap(1) \cap f(p)$ if either $(x,y) \in Ap(1)$ or $(y,z) \in Ap(1)$ by completeness and transitivity of $p(1)$. If neither holds then (x,y) and (y,z) both belong to $Sp(1)$ and hence $(x,z) \in Sp(1)$ because $p(1)$ is a preorder (see 1.22). In that case (x,y), $(y,z) \in f(p)$ implies that (x,y) and (y,z) both belong to $p(2)$ and hence $(x,z) \in p(2)$ because $p(2)$ is a preorder. Therefore, $(x,z) \in f(p)$, and $f(p)$ is also a preorder. (It is continuous because the topology is discrete.) In later chapters every dictatorship will be a complete dictatorship by virtue of the topological and domain assumptions.

Now, consider a different type of social welfare function.

10. Example. $T = \{1,2,3, . . .,n\}$ and X is a discrete space with at least three alternatives. Set $D = P(X)^{\mathrm{T}}$ and $(x,y) \in f(p)$ if and only if $(x,y) \in p(1) \cup p(2)$.

In other words, x ranks above y socially if and only if 1 and 2 both strictly prefer x to y. In this case $\{1,2\}$ is the smallest decisive coalition and both members of the coalition have veto power. We call such a coalition a *proper oligarchy*.

11. Proper oligarchy. Coalition $I \subset T$ is a proper oligarchy if it is globally decisive and every one of its members has veto power. The social welfare function itself is said to be properly oligarchical.

For the domains that we will consider, a decisive coalition that is not minimal cannot be an oligarchy. To illustrate, suppose that I and I' are decisive and I is a proper subset of I'. If there exist x, $y \in X$ and $p \in D$ such that $(x,y) \in \cap_{t \in I} Ap(t)$ but $(y,x) \in Ap(t')$ for some $t' \in I' - I$ then we have $(x,y) \in Af(p)$ because I is decisive, so t' does not have veto power.

Not every properly oligarchical social welfare function defines quasitransitive or even acyclic social preferences. We demonstrate this by means of dictatorship. (If t is a dictator then $\{t\}$ is a proper oligarchy.)

12. Example. $T = \{1\}$ and $X = \{v_1,v_2,v_3\}$ with the discrete topology. Set $D = P(X)$ and $(x,y) \in f(p)$ if and only if (i) $(x,y) \in p(1)$ and (ii) $(x,y) \in Sp(1)$ implies $(x,y) \notin \{(v_3,v_2), (v_2,v_1), (v_1,v_3)\}$.

If (v_1,v_2) and (v_2,v_3) belong to $Sp(1)$ we have (v_1,v_2), $(v_2,v_3) \in Af(p)$, but (v_3,v_1) belongs to $Af(p)$.

An (improper) oligarchy is a group possessing decisiveness and veto power, as in the case of a proper oligarchy, but the preference scheme of one or more individuals is inverted before the definitions of decisiveness and veto power are applied. Such oligarchies are defined by means of an ordered pair (I,J) of subsets of T. We have x ranking above y in the social preference scheme if everyone in I strictly prefers x to y *and* everyone in J strictly prefers y to x. In addition, everyone in I has veto power in the sense of Definition 7 and everyone in J has *inverse* veto power: if individual $j \in J$ strictly prefers x to y then x cannot rank above y in the social preference scheme. Begin with the definition of a decisive *pair* of coalitions.

13. Decisive pairs. If I and J are subsets of T then the ordered pair (I,J) is decisive over (x,y) if $I \cup J \neq \varnothing = I \cap J$ and $(x,y) \in Af(p)$ whenever $(x,y) \in Ap(i)$ for all $i \in I$ and $(y,x) \in Ap(j)$ for all $j \in J$. Then (I,J) is globally decisive (or just decisive) if it is decisive over all ordered pairs.

Note that (I,J) is decisive if and only if $\cap_{i \in I} Ap(i) \cap_{j \in J} -Ap(j)$ is a subset of $Af(p)$ for every profile p. (Of course if I or J is empty we delete the associated part of the intersection operation.)

Just as we have extended the idea of decisiveness to include cases where an individual's preferences are inverted before they are brought to bear on the social preference scheme we now generalize the notion of veto power.

14. Inverse veto power. Individual t has inverse veto power if $-Ap(t) \subset f(p)$ for every profile p.

Now we define oligarchy in the extended sense.

15. Oligarchy. The social welfare function f is oligarchical if there is some pair (I,J) that is decisive for f and such that each member of I has veto power and each member of J has inverse veto power. If $Af(p) = \cap_{i \in I} Ap(i) \cap_{j \in J} -Ap(j)$ for all $p \in D$ then f is completely oligarchical.

16. Proposition. If $f: D \to B(X)$ is completely dictatorial then $f(p)$ is a preorder for each $p \in D$, and if f is completely oligarchical then each $f(p)$ is quasitransitive.

Proof. $D \subset P(X)^T$ so if $f(p) = p(t)$ then $f(p)$ is a preorder. If $Af(p) = \cap_{i \in I} Ap(i) \cap_{j \in J} -Ap(j)$ then $Af(p)$ is transitive because $Ap(i)$ and $-Ap(j)$ are both transitive by Proposition 1.15. ∎

If $(\varnothing, \{t\})$ is an oligarchy we say that t is an inverse dictator and the social welfare function is said to be inversely dictatorial.

17. Inverse dictatorship. If $-Ap(t) \subset Af(p)$ for all $p \in D$ we say that t is an inverse dictator, and the social welfare function is inversely dictatorial. It is completely inversely dictatorial if $f(p) = -p(t)$ for all $p \in D$.

If f is dictatorial or inversely dictatorial we say that it is *authoritarian*, and it is *completely authoritarian* if it is completely dictatorial or completely inversely dictatorial. If f is authoritarian then the preferences of only one individual can influence social preference. A central result is that $f: D \to P(X)$ is constant or authoritarian if X is a connected T_1-space, D has the free-triple property, and f satisfies Arrow's independence axiom (Theorem 8.8). The free triple property holds if for every three-element subset Y of alternatives, all possible profiles of individual preorders on Y are contained in D. (Remember, domains are always assumed to be product sets.)

18. Free-triple property. $D \subset P(X)^T$ has the free-triple property if for every three-element set $\{x,y,z\}$ of alternatives from X we have $D(\{x,y,z\}) = P(\{x,y,z\})^T$.

This property can be extended to m-element subsets of X, where m is some positive integer.

19. Free m-tuple property. $D \subset P(X)^T$ has the free m-tuple property if for every m-element subset Y of X we have $D(Y) = P(Y)^T$.

In the case of infinite societies it will sometimes be necessary to assume that for *every* finite subset Y of X, every profile of preorders on Y is embodied in the domain.

20. Complete domain. D is complete if it has the free m-tuple property for every finite m.

Obviously, every constant or completely authoritarian social welfare function satisfies Arrow's independence axiom. Not every dictatorial (or authoritarian) f satisfies the independence axiom, as we now demonstrate.

21. Example. $T = \{1\}, X = \{v_1, v_2, v_3\}$, $D = P(X)$, and $(x,y) \in f(p)$ if and only if (i) $(x,y) \in p(1)$ and (ii) $(x,y) \in Sp(1)$ implies $(x,y) \neq (v_3, v_2)$ or $(v_1, v_2) \in Sp(1)$.

If individual 1 is indifferent between all three alternatives according to p then $f(p)$ will express universal indifference as well. That is, $(x,y) \in f(p)$ for all $x, y \in X$ if $(x,y) \in p(1)$ for all $x, y \in X$. Suppose, however, that (v_1, v_2) and (v_1, v_3) belong to $Ap'(1)$ and (v_2, v_3) belongs to $Sp'(1)$. Then (v_2, v_3) belongs to $Af(p')$

although (v_3,v_2) belongs to $f(p)$. Although $p(1) \cap \{v_2,v_3\}^2 = p'(1) \cap \{v_2,v_3\}^2$ the social ordering $f(p)$ of v_2 and v_3 is different from the social ordering $f(p')$ of v_2 and v_3. The independence axiom is not satisfied although f is dictatorial.

We conclude this chapter with a discussion of *quasi-oligarchies*, which play a role in some of the lemmas leading to the key results on oligarchies. A quasi-oligarchy is a triple (I,J,K) of coalitions with I and J playing the role of a decisive pair, but in order to force the social ranking of x above y we must verify that the members of K are *indifferent* between x and y.

22. Decisive triples. (I,J,K) is decisive over (x,y) if $I \cup J \neq \emptyset = I \cap J = I \cap K = J \cap K$, and for all $p \in D$ we have $(x,y) \in Af(p)$ if $(x,y) \in Ap(i)$ for all $i \in I$ and $(y,x) \in Ap(j)$ for all $j \in J$ and $(x,y) \in Sp(k)$ for all $k \in K$. The triple (I,J,K) is globally decisive (or just decisive) if it is decisive over all pairs.

The idea of veto power is also extended to quasi-veto power, which is held by an individual for whom indifference between a pair of alternatives is necessary for strict social preference.

23. Quasi-veto power. Individual $t \in T$ has quasi-veto power if we have $(x,y) \in Sf(p)$ whenever $(x,y) \in Ap(t) \cup -Ap(t)$.

A quasi-oligarchy is a triple that is decisive over all pairs and for which each member has the relevant form of veto power.

24. Quasi-oligarchy. The social welfare function f is quasi-oligarchical if there is some decisive triple (I,J,K) such that every member of I has veto power, every member of J has inverse veto power, and every member of K has quasi-veto power. The triple itself is called a quasi-oligarchy.

The members of K have quasi-veto power because neither $(x,y) \in Af(p)$ nor $(y,x) \in Af(p)$ can hold unless each individual $k \in K$ is indifferent between x and y. As one would expect, continuity of social preference implies $K = \emptyset$ in an economic setting: if k belongs to K and $(x,y) \in Af(p)$ then $(x,y) \in Sp(k)$. Continuity of $f(p)$ implies that there is some neighbourhood $N(y)$ of y such that $(x,z) \in Af(p)$ for all $z \in N(y)$. If every neighbourhood of y contains an alternative z such that $(z,y) \in Ap(k)$ then $(z,y) \in Sf(p)$ follows from quasi-veto power. Hence, $f(p)$ cannot be continuous if f is quasi-oligarchical and $K \neq \emptyset$.

We complete the chapter by defining the null social welfare, which sets every outcome socially indifferent to every other outcome in all situations. It is a special case of a constant social welfare function.

25. The null rule. The social welfare function f is null if $f(p) = X \times X$ for all $p \in D$.

Then f is null if $(x,y) \in Sf(p)$ for all $p \in D$ and all $x,y \in X$.

Part II
Discrete Spaces

Introduction

In Part II the topological considerations are ignored, for two reasons. First, this is the setting in which the pathbreaking results were established, Arrow (1951) in particular. Secondly, these preliminary results will be needed in proving stronger impossibility theorems for connected T_1 spaces. In both cases it is necessary to assume that the domain has the free-triple property (3.18). That is, for any three-element subset of alternatives, any profile of individual preorders on that subset is contained in some profile in the overall domain. Clearly, the classical domain W of profiles of selfish and monotonic (etc.) individual preorders on the space Ω of allocations of private goods does not have the free-triple property. For example, choose allocations x, y, and z such that $x(t) \gg y(t) \gg z(t)$ for all $t \in T$. Then every individual strictly prefers x to y and y to z according to every profile p in W. This does not vitiate the proofs of the classical impossibility theorems in Chapters 5 and 6, however. Because $W_t(\{x,y,z\})$, the restriction of W_t to $\{x,y,z\}$, is a singleton when $x(t) \geq y(t) \geq z(t)$ there is no difficulty in finding $p'' \in W$ such that (i) $p''(t) \cap \{x,y\}^2 = p(t) \cap \{x,y\}^2$ and (ii) $p''(t) \cap \{y,z\}^2 = p'(t) \cap \{y,z\}^2$ for arbitrary $p,p' \in W$. The problem arises when $z(t) \geq x(t)$ but $W_t(Z) = P(Z)$ for $Z = \{x,y\}$ and $Z = \{y,z\}$. We may have $(x,y) \in Ap(t)$ and $(y,z) \in Ap'(t)$, but there will be no $p'' \in W$ such that (i) and (ii) both hold because those two statements imply $(x,z) \in Ap''(t)$ in this case. To deal with this problem we will have to work with subsets Y of Ω such that every triple is arbitrarily close (in a topological sense) to a free triple. We then extend impossibility theorems for $f \mid Y$, the restriction of the social welfare function to Y, to a single, global impossibility theorem. The impossibility theorems for $f \mid Y$ are themselves proved by 'integrating' results on discrete subspaces of Y. Any finite subset of Y will be a discrete space with respect to the relative topology, and the subdomain will have the free-triple property. Let us begin, then, with discrete spaces.

4 Decisiveness Lemmas

IF the social welfare function f is non-null then for some pair of alternatives x and y, and some profile p in the domain, we have $(x,y) \in Af(p)$. (Recall that AR is the asymmetric factor of the binary relation R and $-R$ denotes the inverse of R. See 1.5, 1.35, and 3.25.) If the sets $I = \{ t \in T \mid (x,y) \in Ap(t) \}$ and $J = \{ t \in T \mid (y,x) \in Ap(t) \}$, consisting of (respectively) the individuals who strictly prefer x to y and those strictly preferring y to x, comprise a partition of T then (I,J) is decisive for the pair (x,y) if f satisfies Arrow's independence axiom. That is, if the independence axiom holds are there are two alternatives x and y such that for profile p there is no individual or social indifference between x and y then the pair of coalitions (I,J) is decisive for at least that ordered pair. In this chapter we show how to prove that (I,J) is actually decisive over all pairs, under some mild conditions on f. Some related results are proved as well, and these constitute a set of lemmas from which a variety of impossibility theorems can be derived with relative ease. Our method of proof is essentially the *modus operandi* of Sen (1986*a*) which provides an elegant and insightful proof of Arrow's impossibility theorem. Once the basic decisiveness results are available, the proofs of the impossibility theorems themselves are almost complete.

All the results in this chapter pertain to a discrete space X of alternatives, to a domain $D \subset P(X)^T$, and to a social welfare function f on D. In each case, social preference will be at least quasitransitive. Continuity of preference plays no role because X is discrete. The fact that X itself may be contained in some larger space has no bearing on the logic of the proofs; simply assume some set X of alternatives along with a domain D of profiles of individual preorders on X. T can be any set in the case of the first series of lemmas. In particular, it is not necessary to assume that T is finite. The last three lemmas do require the finiteness assumption, however.

Before turning to the formal arguments we highlight the role played by the free-triple property. The domain D is assumed to be a product set, $D = \Pi_{t\in T}D_t$. Let x, y, and z be any three distinct members of X, and suppose that $D(\{x,y,z\}) = P(\{x,y,z\})^T$. Now, let p and p' be any two profiles in D. Each time we apply the free-triple hypothesis it is to justify the assumption that there is a profile p'' such that, for each $t \in T$, $p''(t)$ and $p(t)$ agree with respect to the ordering of x and y, and $p''(t)$ and $p'(t)$ agree with respect to the ordering of y and z. Let us see why this is so. Let R be any complete order on $\{x,y\}$ and let R' be any complete order on $\{y,z\}$. It is easy to see that R and R' can never be in conflict. Whatever the specification of R and R' there will always be some *preorder* R'' on $\{x,y,z\}$ such that $R'' \cap \{x,y\}^2 = R$ and $R'' \cap \{y,z\}^2 = R'$. Therefore, if $R = p(t) \cap \{x,y\}^2$ and $R' = p'(t) \cap \{y,z\}^2$, any ordering $p''(t) \in D_t$ such that $p''(t) \cap \{x,y,z\}^2 = R''$ will meet our requirement. Because R'' is a preorder and $\{x,y,z\}$ is a triple, the free-triple property guarantees the existence of $p''(t)$ in D_t.

Some new terminology is required at this stage.

1. Definition. If $x,y \in X$ and $(x,y) \in Af(p)$ holds for some $p \in D$ we write $x >_f y$ to represent that fact. If $x,y \in X$ and $(x,y) \in f(p)$ holds for some $p \in D$ we write $x \geq_f y$.

With this new notation, and the understanding that *in this chapter* only domains in which all pairs of alternatives are a priori unrestricted are considered, the non-imposition condition is

$x \geq_f y$ for all $x,y \in X$

and the strict non-imposition condition is

$x >_f y$ for all $x,y \in X$ such that $x \neq y$.

(Recall definitions 3.3 and 3.2.)

Now we are ready for the basic lemmas on the power structure underlying a social welfare function. The first result is a proposition in logic that is used as a concluding step in many of the subsequent lemmas. It is useful to know that all ordered pairs have property C if there is a single ordered pair (x,y) with property C *and* we can prove that (x,z) and (z,y) satisfy C for any z whenever (x,y) satisfies C.

2. Lemma. Suppose that X has at least three alternatives and $R \neq \varnothing$ is an irreflexive binary relation on X. If for any choice of distinct $x,y,z \in X$ we have (x,z), $(z,y) \in R$, whenever $(x,y) \in R$, then (x,y) belongs to R for all $x,y \in X$ such that $x \neq y$.

Proof. Consider the following two statements:

2a $x \neq y \neq z \neq x$ and $(x,y) \in R$ imply $(x,z) \in R$.
2b $x \neq y \neq z \neq x$ and $(x,y) \in R$ imply $(z,y) \in R$.

These statements are part of the hypothesis. Now, assume that $(v,w) \in R$. Choose any $z \in X - \{v,w\}$. We can easily establish

2c $(x,y) \in R$ for all $x,y \in \{v,w,z\}$ such that $x \neq y$.

This is accomplished by the following:

$$(v,w) \to 2a \to (v,z) \to 2b \to (w,z) \to 2a \to (w,v) \to 2b \to (z,v)$$
$$(v,w) \to 2b \to (z,w).$$

The statement $(r,s) \to 2n \to (c,d)$ means that property $2n$ is used to establish $(c,d) \in R$ given that (r,s) belongs to R.

Now choose r and s such that $r \neq s$. If $\{r,s\} \cap \{v,w\} \neq \emptyset$ choose $z \in X - \{r,s\}$. Because $(v,w) \in R$ and 2a and 2b hold we have $(x,y) \in R$ for some $x,y \in \{r,s,z\}$ such that $x \neq y$. Then $(r,s),(s,r) \in R$ by 2c. If $\{r,s\} \cap \{v,w\} = \emptyset$ then $(r,s),(s,r) \in R$ by 2c applied to $\{r,s,v\}$ because $(v,r) \in R$ holds by 2a and the fact that $(v,w) \in R$. ∎

Now we begin the series of basic lemmas on the power structure of a social welfare function satisfying Arrow's independence axiom. With the exception of the next result, Lemma 3, they all assume strict non-imposition and quasitransitive social preference. Lemma 3 merely shows that f satisfies strict non-imposition if it is non-null (3.25) and social preference is fully transitive. Therefore, all of the subsequent lemmas apply either to a function $f: D \to Q(X)$ satisfying strict non-imposition, or to a non-null $f: D \to P(X)$ satisfying non-imposition.

3. Lemma. Let T be any set. Suppose that $f: D \to P(X)$ is a non-null social welfare function satisfying non-imposition and Arrow's independence axiom. If X has at least three alternatives and D has the free-triple property then f satisfies strict non-imposition.

Proof. Because f is non-null we have $(x,y) \in Af(p)$ for some $p \in D$ and $x,y \in X$. We begin by proving the following statement, which corresponds to 2a.

3a If $x \neq z \neq y$ then $x >_f z$.

We have $(y,z) \in f(p')$ for some $p' \in D$ because f satisfies non-imposition. Because D has the free-triple property, there is some $p'' \in D$ such that, for each $t \in T$, $p''(t) \cap \{x,y\}^2 = p(t) \cap \{x,y\}^2$ and $p''(t) \cap \{y,z\}^2 = p'(t) \cap \{y,z\}^2$. Then $(x,y) \in Af(p'')$

and $(y,z) \in f(p'')$ by the independence axiom. Then $(x,z) \in Af(p'')$ because $f(p'')$ is complete and transitive. (Recall 1.27.) Now, prove the following counterpart to 2b.

3b If $x \neq z \neq y$ then $z >_f y$.

By non-imposition, $(z,x) \in f(p')$ for some $p' \in D$. Choose $p'' \in D$ so that, for each $t, p''(t)$ and $p(t)$ agree on $\{x,y\}$ and $p''(t)$ and $p'(t)$ agree on $\{x,z\}$. Then $(z,x) \in f(p'')$ and $(x,y) \in Af(p'')$ by independence, and thus $(z,y) \in Af(p'')$ because $f(p'')$ is a preorder.

Now we have $x >_f y$ for all $x,y \in X$ such that $x \neq y$ by Lemma 2 applied to the irreflexive relation $>_f$. ∎

Next we prove the critical 'extension lemma': If a pair of coalitions is decisive for some pair of distinct alternatives then it will be globally decisive. Recall the definition of decisiveness, 3.13.

4. Lemma. Suppose that $f: D \to Q(X)$ satisfies Arrow's independence axiom and strict non-imposition. If X has at least three members and D has the free-triple property then a pair (I,J) of coalitions is globally decisive if it is decisive for some pair of distinct alternatives.

Proof. Suppose that (I,J) is decisive over (x,y) and $x \neq y$. First we prove:

4a If $x \neq z \neq y$ then (I,J) is decisive over (x,z).

Let p be any profile for which $(y,z) \in Af(p)$. Choose $p' \in D$ arbitrarily except that, for each $t \in T$, $p'(t) \cap \{y,z\}^2 = p(t) \cap \{y,z\}^2$ and $(x,y) \in Ap'(i)$ for all $i \in I$, $(y,x) \in Ap'(j)$ for all $j \in J$. This is possible by the free-triple assumption. Then we have $(x,y) \in Af(p')$ by decisiveness of (I,J) over (x,y), and $(y,z) \in Af(p')$ by the independence axiom. Then $(x,z) \in Af(p')$ by transitivity of $Af(p')$. Notice that the definition of p' is consistent with any ordering of x and z for $t \notin I \cup J$ and is consistent with $(x,z) \in \cap_{i \in I} Ap(i) \cap_{j \in J} - Ap(j)$. Therefore, (I,J) is decisive over (x,z).

Similarly, we establish

4b If $x \neq z \neq y$ then (I,J) is decisive over (z,y).

We have $(z,x) \in Af(p)$ for some $p \in D$ by strict non-imposition. Choose $p' \in D$ arbitrarily except that $p'(t) \cap \{z,x\}^2 = p(t) \cap \{z,x\}^2$ for each $t \in T$, and $(x,y) \in \cap_{i \in I} Ap'(i) \cap_{j \in J} - Ap'(j)$.

Then $(z,x) \in Af(p')$ by independence and $(x,y) \in Af(p')$ by decisiveness of (I,J) over (x,y). Therefore, $(z,y) \in Af(p')$ by transitivity of $Af(p')$ and thus (I,J) is decisive over (z,y).

Global decisiveness of (I,J) follows from Lemma 2 if R is defined by setting $(x,y) \in R$ whenever $x \neq y$ and (I,J) is decisive for (x,y). ∎

We have just shown that decisiveness over a pair extends to decisiveness over all pairs. Now we show that decisiveness is preserved by component-wise intersection.

5. Lemma. Suppose that $f: D \to Q(X)$ satisfies Arrow's independence axiom and strict non-imposition. If X has at least three members and D has the free-triple property then $(I \cap I', J \cap J')$ is decisive if (I,J) and (I',J') are both decisive.

Proof. Suppose that (I,J) and (I',J') are both decisive. Let x, y, and z be any three distinct members of X. Choose $p \in D$ arbitrarily except that:

$(x,y) \in Ap(t)$ if $t \in I$.
$(y,z) \in Ap(t)$ if $t \in I'$.
$(y,x) \in Ap(t)$ if $t \in J$.
$(z,y) \in Ap(t)$ if $t \in J'$.

We have $(x,y) \in Af(p)$ because (I,J) is decisive, and $(y,z) \in Af(p)$ because (I',J') is decisive. Then $(x,z) \in Af(p)$ because $Af(p)$ is transitive. Then $(I \cap I', J \cap J')$ is decisive for (x,z) because the choice of p is consistent with any ordering of x and z for an individual t not belonging to $I \cap I'$ or to $J \cap J'$. Therefore, $(I \cap I', J \cap J')$ is decisive by Lemma 4. ∎

We still have not proved that decisive pairs of coalitions exist. If social preference is fully transitive then decisive pairs will exist, whether T is finite or not. If social preference is merely quasitransitive then decisive pairs may not exist even if T is finite. This is demonstrated by Example 6.

6. Example. X is any set of three or more members, $T = \{1,2\}$, and $D = P(X)^T$. Set $(x,y) \in f(p)$ if and only if $(x,y) \in p(1)$ or $(x,y) \in Ap(2)$ or $(y,x) \in Ap(2)$.

In this case each $f(p)$ is quasitransitive. We have $(x,y) \in Af(p)$ if and only if $(x,y) \in Ap(1)$ and $(x,y) \in Sp(2)$. Then quasitransitivity of $f(p)$ follows from transitivity of the relations $Ap(1)$ and

Sp (2). Obviously, *f* satisfies strict non-imposition and the independence axiom. But there are no decisive pairs of coalitions.

What we can prove at this stage is that if *T* is finite and a decisive pair exists then there exists a *minimal decisive pair*.

7. Minimal decisive pair.

A pair of coalitions (I, J) is a minimal decisive pair (with respect to a given social welfare function) if it is a decisive pair and for every decisive pair (I', J') we have $I \subset I'$ and $J \subset J'$.

If *T* is finite then the set $\{ (I^1, J^1), (I^2, J^2), \ldots, (I^m, J^m) \}$ of all decisive pairs is finite. Therefore, $(\cap I^h, \cap J^h)$ is a minimal decisive pair by Lemma 5 *if* the set of decisive pairs is non-empty. (Of course, *h* varies from 1 to *m* as intersection proceeds.) We continue to postpone the issue of existence of a decisive pair, and we next prove that the members of a minimal decisive pair have veto power and inverse veto power.

8. Lemma.

Suppose that $f: D \to Q(X)$ satisfies Arrow's independence axiom and strict non-imposition. If *X* has at least three members, *D* has the free-triple property, and (I^*, J^*) is a minimal decisive pair then every member of I^* has veto power and every member of J^* has inverse veto power.

Proof. Suppose that t^* belongs to I^* but t^* does not have veto power. Then we can find $x, y \in X$ and $p \in D$ such that $(x, y) \in Ap(t^*)$ but $(y, x) \in Af(p)$. Choose any $z \in X - \{x, y\}$. Now choose $p' \in D$ arbitrarily except that

$(z, y) \in Ap'(t)$ if $t \in I^*$,

$(y, z) \in Ap'(t)$ if $t \in J^*$,

and $p'(t) \cap \{x, y\}^2 = p(t) \cap \{x, y\}^2 \; \forall \, t \in T$. Then $(z, y) \in Af(p')$ by decisiveness of (I^*, J^*), and $(y, x) \in Af(p')$ by the independence axiom. Therefore, $(z, x) \in Af(p')$ by transitivity of $Af(p')$. Our choice of p' is consistent with any ordering of *z* and *x* by an individual *t* not belonging to $I^* - \{t^*\}$ or to J^*. Therefore, $(I^* - \{t^*\}, J^*)$ is decisive for (z, x) and hence is globally decisive by Lemma 4. This contradicts the fact that (I^*, J^*) is minimal. Therefore, each member of I^* has veto power.

Suppose now that t^* belongs to J^* and for some $x, y \in X$, $p \in D$ we have $(x, y) \in Af(p) \cap Ap(t^*)$. Choose $z \in X - \{x, y\}$, and $p' \in D$ arbitrarily except that

$(y,z) \in Ap'(t)$ if $t \in I^*$,
$(z,y) \in Ap'(t)$ if $t \in J^*$,

and $p'(t) \cap \{x,y\}^2 = p(t) \cap \{x,y\}^2$ for all $t \in T$. Then (x,y), $(y,z) \in Af(p')$ by independence and decisiveness of (I^*,J^*), respectively. Then $(x,z) \in Af(p')$ by transitivity of $Af(p')$. Our choice of p' is consistent with any ordering of x and z by individuals not belonging to I^* or $J^* - \{t^*\}$. Therefore, $(I^*,J^* - \{t^*\})$ is decisive over (x,z) and thus over all pairs by Lemma 4. This contradicts the fact that (I^*,J^*) is a minimal decisive pair. Therefore, the members of J^* have inverse veto power. ∎

The proofs of the basic impossibility theorems of Arrow (1963) and Wilson (1972) for fully transitive social preference will employ the lemmas that have just been proved. The remaining results are used to establish the oligarchy theorems in the case of quasi-transitive social preference. Although a decisive pair need not exist in that setting there will always be a decisive *triple* (I,J,K) of coalitions, as we proceed to demonstrate (Definition 3.22). It is necessary to assume that T is finite however. The existence theorem for infinite T (Chapter 9) depends on continuity of social preference. Lemmas 9 and 12 generalize Lemmas 4 and 5 but we provide separate proofs in order to have a direct path to the Arrow and Wilson theorems.

9. Lemma. Suppose that $f: D \to Q(X)$ satisfies Arrow's independence axiom and strict non-imposition, X has at least three members, and D has the free-triple property. If (I,J,K) is decisive for some pair of distinct alternatives, and for every triple (I',J',K') that is decisive for some distinct pair we have $K = K'$ if $K' \subset K$ then (I,J,K) is globally decisive.

Proof. Suppose that (I,J,K) is decisive for (x,y) and $x \neq y$. We begin by proving

9a If $x \neq z \neq y$ then (I,J,K) is decisive for (x,z).

Choose $p \in D$ such that $(y,z) \in Af(p)$. Set $K_1 = \{t \in K \mid (y,z) \in Ap(t)\}$, $K_2 = \{t \in K \mid (z,y) \in Ap(t)\}$, and $K_3 = \{t \in K \mid (y,z) \in Sp(t)\}$. Now choose $p' \in D$ arbitrarily except that

$(x,y),(x,z) \in Ap'(t)$ if $t \in I$.
$(y,x),(z,x) \in Ap'(t)$ if $t \in J$.
$(x,y) \in Sp'(t)$ if $t \in K$.
$p'(t) \cap \{y,z\}^2 = p(t) \cap \{y,z\}^2$ for all $t \in T$.

Then $(x,y) \in Af(p')$ by decisiveness of (I,J,K) over (x,y). And $(y,z) \in Af(p')$ by the independence axiom. Therefore $(x,z) \in Af(p')$ by transitivity of $Af(p')$. The choice of $p' \in D$ is consistent with any ordering of x and z by any t not belonging to $I \cup J \cup K$. Therefore, $(I \cup K_1, J \cup K_2, K_3)$ is decisive for (x,z) by independence. But $K_3 \subset K$ by definition, so $K = K_3$ by hypothesis. Therefore $K_1 = \varnothing = K_2$. Therefore, (I,J,K) is decisive for (x,z).

Now prove

9b If $x \neq z \neq y$ then (I,J,K) is decisive for (z,y).

Choose $p \in D$ such that $(z,x) \in Af(p)$. Set $K_1 = \{t \in K \mid (z,x) \in Ap(t)\}$, $K_2 = \{t \in K \mid (x,z) \in Ap(t)\}$, and $K_3 = \{t \in K \mid (z,x) \in Sp(t)\}$. Choose $p' \in D$ arbitrarily except that

(x,y), $(z,y) \in Ap(t)$ if $t \in I$.
(y,x), $(y,z) \in Ap(t)$ if $t \in J$.
$(x,y) \in Sp(t)$ if $t \in K$.
$p'(t) \cap \{z,x\}^2 = p(t) \cap \{z,x\}^2$ for all $t \in T$.

Then $(z,x) \in Af(p')$ by independence and $(x,y) \in Af(p')$ by decisiveness of (I,J,K) over (x,y). Then $(z,y) \in Af(p')$ and thus $(I \cup K_1, J \cup K_2, K_3)$ is decisive for (z,y) and thus $K_3 \subset K \subset K_3$. Therefore, (I,J,K) is decisive for (z,y).

We have established 9a and 9b and thus (I,J,K) is globally decisive by Lemma 2. ∎

Another definition is required at this point.

10. Proper decisiveness. (I,J,K) is a proper decisive triple if it is globally decisive and for any triple (I',J',K') that is decisive for some distinct pair of outcomes we have $K = K'$ if $K' \subset K$. We say that K is minimal in this case.

11. Lemma. Assume that T is finite. Suppose that $f: D \to Q(X)$ satisfies Arrow's independence axiom and strict non-imposition, X has at least three members, and D has the free-triple property. Then there exists a proper decisive triple.

Proof. Choose any distinct x and y in X and $p \in D$ such that $(x,y) \in Af(p)$. If $(x,y) \in Sp(t)$ for all $t \in T$ then choose $p' \in D$ such that $(y,x) \in Af(p')$. Then we must have $(y,x) \in Ap(t) \cup -Ap(t)$ for some $t \in T$ by Arrow's independence axiom. Therefore, there exist $x,y \in X$ and $p \in D$ such that $I \cup J \neq \varnothing$ for $I = \{t \in T \mid (x,y) \in Ap(t)\}$ and $J = \{t \in T \mid (y,x) \in Ap(t)\}$. Set

$K = \{t \in T \mid (x,y) \in Sp(t)\}$. Then (I,J,K) is decisive for (x,y) by the independence axiom. Therefore, the set Σ of triples (I',J',K') such that (I',J',K') is decisive for some pair of alternative is not empty. The set is finite because T is finite. Then three exists a triple (I',J',K') in Σ such that $|K'| \leq |K|$ for all $(I,J,K) \in \Sigma$. (The vertical bars denote cardinality.) Then (I',J',K') is a proper decisive triple by Lemma 9. ∎

12. Lemma. Assume that T is finite. Suppose that $f: D \to Q(X)$ satisfies Arrow's independence axiom and strict non-imposition, X has at least three members, and D has the free-triple property. If (I,J,K) and (I',J',K') are two proper decisive triples then $(I \cap I', J \cap J', K \cap K')$ is a proper decisive triple.

Proof. Let x, y, and z be any three distinct members of X. Choose $p \in D$ arbitrarily, except that

$(x,y) \in Ap(t)$ if $t \in I$.
$(y,z) \in Ap(t)$ if $t \in I'$.
$(y,x) \in Ap(t)$ if $t \in J$.
$(z,y) \in Ap(t)$ if $t \in J'$.
$(x,y) \in Sp(t)$ if $t \in K$.
$(y,z) \in Sp(t)$ if $t \in K'$.

Then $(x,y) \in Af(p)$ by decisiveness of (I,J,K), and $(y,z) \in Af(p)$ by decisiveness of (I',J',K'). Then $(x,z) \in Af(p)$ by transitivity of $Af(p)$. Because the choice of p is consistent with any ordering of x and z for any individual t not belonging to $I \cap I'$, or to $J \cap J'$, or to $K \cap K'$, the triple $(I \cap I', J \cap J', K \cap K')$ is decisive over (x,z) by the independence axiom. Then it is globally decisive by Lemma 9. $K \cap K'$ is minimal because K and K' are both minimal. We must have $K = K'$ in fact. ∎

A *minimal decisive triple* is of course a decisive triple (I^*,J^*,K^*) such that $I^* \subset I, J^* \subset J$, and $K^* \subset K$ for any decisive triple (I,J,K).

If T is finite then a minimal decisive triple exists by Lemmas 11 and 12. This minimal decisive triple must be unique (by Lemma 12). We conclude this chapter by proving that each member of such a triple has a form of veto power.

13. Lemma. Assume that T is finite. Suppose that $f: D \to Q(X)$ satisfies Arrow's independence axiom and strict non-imposition, X

has at least three members, and D has the free-triple property. Then there exists a minimal decisive triple (I^*,J^*,K^*) and each member of I^* has veto power, each member of J^* has inverse veto power, and each member of K^* has quasi-veto power.

Proof. We have already remarked that existence is established by Lemmas 11 and 12.

Suppose that $t^* \in I^*$, $x,y \in X$, $p \in D$, and $(x,y) \in Ap(t^*) \cap -Af(p)$. Choose any $z \in X - \{x,y\}$ and then $p' \in D$ arbitrarily except that

$(z,y) \in Ap'(t)$ if $t \in I^*$.

$(y,z) \in Ap'(t)$ if $t \in J^*$.

$(z,y) \in Sp'(t)$ if $t \in K^*$.

$p'(t) \cap \{x,y\}^2 = p(t) \cap \{x,y\}^2$ for all $t \in T$.

Then $(z,y) \in Af(p')$ by decisiveness of (I^*,J^*,K^*) and $(y,x) \in Af(p')$ by the independence axiom. Then $(z,x) \in Af(p')$ by transitivity of $Af(p')$. Then the triple $(I^* - \{t^*\},J^*,K^*)$ is decisive for (z,x) because the choice of p' is consistent with any ordering of x and z for t not belonging to $I^* - \{t^*\}$, J^*, or K^*. Obviously, K^* is minimal. (Use the argument of Lemma 12.) Therefore, $(I^* - \{t^*\},J^*,K^*)$ is decisive by Lemma 9. This contradicts the definition of (I^*,J^*,K^*). Therefore, each member of I^* must have veto power.

Suppose $t^* \in J^*$ and $(x,y) \in Ap(t^*) \cap Af(p)$. Choose $z \in X - \{x,y\}$ and $p' \in D$ satisfying

$(y,z) \in Ap'(t)$ if $t \in I^*$.

$(z,y) \in Ap'(t)$ if $t \in J^*$.

$(y,z) \in Sp'(t)$ if $t \in K^*$.

$p'(t) \cap \{x,y\}^2 = p(t) \cap \{x,y\}^2$ for all $t \in T$.

Then $(x,y),(y,z) \in Af(p')$ by independence and decisiveness of (I^*,J^*,K^*), respectively. Then $(x,z) \in Af(p')$ and thus $(I^*,J^* - \{t^*\},K)$ is decisive for (x,z) because the choice of p' is consistent with any ordering of x and z by t not belonging to I^* or $J^* - \{t^*\}$ or K^*. Therefore, $(I^*,J^* - \{t^*\},K^*)$ is decisive by Lemma 9. This contradiction establishes that each member of J^* has inverse veto power.

Finally, suppose that t^* belongs to K^* and (x,y) belongs to both $Af(p)$ and $Ap(t^*) \cup -Ap(t^*)$. Set $I' = \{t \in T \mid (x,y) \in Ap(t)\}$, $J' = \{t \in T \mid (y,x) \in Ap(t)\}$, and $K' = \{t \in T \mid (x,y) \in Sp(t)\}$. Then (I',J',K') is decisive for (x,y) by the independence axiom.

Because T is finite there is some proper decisive triple (I'', J'', K'') such that $K'' \subset K'$ (Lemma 9). But $t^* \notin K'$ so $t^* \notin K''$. But this contradicts the definition of (I^*, J^*, K^*). Therefore, each member of K^* has a quasi-veto. ∎

Note that $(\{1\}, \varnothing, \{2\})$ is a decisive triple for the social welfare function of Example 6.

5 Transitive Social Preference

THIS chapter proves K. J. Arrow's impossibility theorem, which established a new field of inquiry in philosophy and the social sciences (Arrow 1951, 1963). We also prove Wilson's generalization of Arrow's theorem (Wilson 1972). Wilson's was the first impossibility theorem to clarify the role played by the Pareto criterion in Arrow's proof and to show how that condition could be substantially weakened. This paved the way for the strong results on the impossibility of equity–efficiency trade-offs in Campbell (1990a,b; 1991$a–d$; 1992) which are presented in Chapters 8–13. In this chapter we also give Fishburn's demonstration that the Arrow and Wilson theorems do not go through for an infinite society and a discrete outcome space (Fishburn 1970).

The proofs of the Arrow and Wilson theorems are based on Lemmas 4.3, 4.4, and 4.5. In words, a pair of coalitions that is decisive over at least two alternatives is globally decisive, and if two pairs of coalitions are decisive then so is the pair obtained by component-wise intersection. Recall that a subset I of T is a decisive *coalition* if $(I,\varnothing)$ is decisive (3.6). Similarly, refer to J as an inversely decisive *coalition* if $(\varnothing,J)$ is decisive. We will show that T is either decisive or inversely decisive and from this it will follow that there is an individual who is either a dictator or an inverse dictator. The assumption that T is finite does not have to be made until the last stage. The hypothesis requires a non-null social welfare function f satisfying Arrow's independence axiom and a domain with the free-triple property. If the Pareto criterion is also satisfied then f cannot be inversely dictatorial so it must be dictatorial. Any topology consistent with the free-triple property may be assumed for X.

1. Lemma. Assume that X has at least three alternatives and D has the free-triple property. If $f: D \to P(X)$ is a non-null social welfare function satisfying non-imposition and Arrow's independence axiom then T is either decisive or inversely decisive.

Proof. If f is non-null then $(x,y) \in Af(p)$ for some $x,y \in X$ and $p \in D$. Choose $z \in X - \{x,y\}$ and $p' \in D$ arbitrarily except that $p'(t) \cap \{x,y\}^2 = p(t) \cap \{x,y\}^2$ for all $t \in T$, $(x,z) \in Ap'(t)$ for all $t \in T$, and $(y,z) \in Ap'(t)$ for all $t \in T$. We have $(x,y) \in Af(p')$ by Arrow's independence axiom. If $(y,z) \in f(p')$ then $(x,z) \in Af(p')$ because $f(p')$ is complete and transitive (1.27). This implies that T is decisive for (x,z) by the independence axiom and hence that T is decisive for all pairs by Lemmas 4.3 and 4.4. If $(z,y) \in Af(p')$ then T is inversely decisive over (y,z) by the independence axiom and hence T is inversely decisive by Lemmas 4.3 and 4.4. ■

Now we prove that if a coalition is neither decisive nor inversely decisive then its complement must be either decisive or inversely decisive.

2. Lemma. Assume that X has at least three alternatives, D has the free-triple property, and T is decisive for f. If $f: D \to P(X)$ satisfies non-imposition and Arrow's independence axiom then for all $I \subset T$ either I or $T - I$ is decisive.

Proof. Choose any three distinct alternatives x, y, and z. Choose any $p \in D$ such that $(x,y) \in Ap(t)$ for all $t \in T$, $(x,z) \in Ap(t)$ for all $t \in I$, and $(z,y) \in Ap(t)$ for all $t \notin I$. Then $(x,y) \in Af(p)$ because T is decisive. If $(z,x) \in f(p)$ then $(z,y) \in Af(p)$ because $f(p)$ is a preorder. Then $T - I$ is decisive for (z,y) by the independence axiom because p is consistent with any ordering of y and z by individuals in I. In that case $T - I$ is decisive for all pairs by Lemmas 4.3 and 4.4. If $(x,z) \in Af(p)$ then I is decisive for (x,z) by the independence axiom and the fact that p is consistent with any ordering of x and z by individuals not in I. Therefore I is decisive for all pairs by Lemmas 4.3 and 4.4. ■

The corresponding result holds for inverse decisiveness.

3. Lemma. Assume that X has at least three alternatives, D has the free-triple property, and T is inversely decisive for f. If $f: D \to P(X)$ satisfies non-imposition and Arrow's independence axiom then for all $I \subset T$ either coalition I or coalition $T - I$ is inversely decisive.

Proof. Let p be specified as in the proof of Lemma 2. We have $(y,x) \in Af(p)$ because T is inversely decisive. If $(x,z) \in f(p)$ then $(y,z) \in Af(p)$ and hence $T - I$ is inversely decisive by the independence axiom and Lemmas 4.3 and 4.4. If $(z,x) \in Af(p)$ then I is inversely decisive. ∎

Now we can prove a fundamental impossibility theorem for finite societies.

4. Theorem: (R.B. Wilson). Assume that T is finite, X has at least three alternatives, and D has the free-triple property. If $f: D \rightarrow P(X)$ is a non-null social welfare function satisfying non-imposition and Arrow's independence axiom then it is either dictatorial or inversely dictatorial.

Proof. Suppose that $T = \{1,2, \ldots ,n\}$ is decisive. Set $T_t = \{i \in T \,|\, i \neq t\}$. If t is not a dictator then coalition T_t is decisive by Lemma 2. If T_t is decisive for $t = 1,2, \ldots ,n - 1$ then $\{n\} = \cap \{T_t \,|\, t = 1,2, \ldots ,n - 1\}$ is decisive by Lemmas 4.3 and 4.5 applied $n - 2$ times. Therefore, if individual t is not a dictator for $t \leq n - 1$ then individual n is a dictator.

If T is not decisive then it is inversely decisive by Lemma 1. If t is not an inverse dictator for $t \leq n - 1$ then $\{n\} = \cap \{T_t \,|\, t = 1,2, \ldots ,n - 1\}$ is inversely decisive by Lemma 3 and Lemmas 4.3 and 4.5, and thus n is an inverse dictator. ∎

This result immediately leads to the famous Arrow theorem.

5. Theorem: (K.J. Arrow). Assume that T is finite, X has at least three alternatives, and D has the free-triple property. If $f: D \rightarrow (X)$ satisfies the Pareto criterion and Arrow's independence axiom then f is dictatorial.

Proof. T is decisive by the Pareto criterion. Therefore, f cannot be inversely dictatorial or null. Therefore, f is dictatorial by Theorem 4. ∎

The domain $P(X)^T$ has the free-triple property if the topology is discrete, and hence Theorems 4 and 5 are valid for $D = P(X)^T$, the case actually considered by Arrow and Wilson. The fact that $P(X)^T$ has the free-triple property is obvious if X is finite and the topology is discrete. If X is infinite we can still prove that any preorder on an arbitrary triple is contained in some preorder on X: Suppose that W, Y, and Z are three non-empty and mutually disjoint subsets of X. Define the preorder R on X by setting $(x,y) \in R$ if and only if (1) $x \in Y$ implies $y \notin W$ and (2) $x \notin W \cup Y$ implies $y \notin W \cup Y$. Then $(x,y) \in SR$ if x and y both belong to W or they both belong to Y or they both belong to Z. This proves that $P(X)^T$

has the free-triple property if $P(X)$ contains all the preorders on X. It is sometimes useful to state the theorems for $D = L(X)^T$ where $L(X)$ is the set of linear orders on X. (A preorder $R \in P(X)$ belongs to $L(X)$ if and only if $(x,y) \in SR$ implies $x = y$.) Obviously, $L(\{x,y,z\}) \neq P(\{x,y,z\})$ if the alternatives in question are distinct. Nevertheless, the proofs underlying Theorems 4 and 5 yield the corresponding impossibility theorems for $D = L(X)^T$. For any choice of $p,p' \in L(X)^T$ any construction required by the proofs and leading to a new profile p'' can be carried out without violating the requirement that p'' belongs to $L(X)^T$. But we have to appeal to Proposition 1.37 for assurance that every linear order on a triple from X is contained in some *linear* order on X, and for this result it is necessary to assume Zorn's Lemma (1.36) unless X is finite. We state more general result as Theorem 6.

6. Theorem. Assume that T is finite and X is a discrete space with at least three alternatives. Suppose that $f: D \to P(X)$ satisfies non-imposition and Arrow's independence axiom. If $D(Y) = P(Y)^T$ or $D(Y) = L(Y)^T$ for every three-element subset Y of X then f is authoritarian if it is not null, and it is dictatorial if it satisfies the Pareto criterion.

Notice that we have weakened the hypothesis beyond merely requiring $D = P(X)^T$ or $D = L(X)^T$. Because the proofs only make use of the free-triple property, they are valid if for an arbitrary three-element subset Y *either* all the profiles of preorders on Y are included in the domain *or* the restriction of the domain to Y coincides with the set of all profiles of linear orders on Y. To demonstrate that this is weaker than requiring $D = P(X)^T$ or $D = L(X)^T$ consider the case $X = \{w,x,y,z\}$ endowed with the discrete topology. If for each t the set D_t consists of all linear orders R on X such that $(v,w) \in AR$ for some $v \in X$ then $D_t(Y)$ contains all the linear orders on Y for every *proper* subset Y of X. It would be false, however, to say that the hypothesis of Theorem 6 allowed $D(Y) = P(Y)^T$ for some triple Y and $D(Y') = L(Y')^T$ for some triple Y'. It is easy to see why. Set $(x,y) \in R_t$ if and only if $(x,y) \in Sp(t)$ for some $p \in D$ and some $t \in T$. If $(x,y) \in R_t$ for some distinct $x,y \in X$ then $(x,z) \in R_t$ for all $z \in X$ because $D(\{x,y,z\}) = P(\{x,y,z\})^T$ or $D(\{x,y,z\}) = L(\{x,y,z\})^T$. We can use this fact to establish $(a,b) \in R_t$ for all $a,b \in X$. Therefore, if $D_t(Y) = P(Y)$ for some three-element set Y and it is assumed that $D(Y) = P(Y)^T$ or $D(Y) = L(Y)^T$ holds for all three-element sets then the former must hold in fact. (See also Schmitz (1977).)

Neither Theorem 4 nor Theorem 5 goes through with every product domain D satisfying $L(X)^T \subset D \subset P(X)^T$, even if X has only three members. Consider the following example taken from Bordes and Le Breton (1990).

7. Example. $X = \{x,y,z\}$, a three-alternative set, and $T = \{1,2,\ldots,n\}$ with $n \geq 2$. Let R^0 denote the null relation. $R^0 = \{(a,b)\,|\,a,b \in X\}$. Set $D_t = L(X) \cup \{R^0\}$ for each $t \in T$, where $L(X)$ is the family of linear orders on X. For any $p \in D$ set $f(p) = p(2)$ if $p(t) = R^0$ for some $t \in T$ and set $f(p) = p(1)$ otherwise.

This social welfare function is not dictatorial because $(x,y) \notin Af(p)$ if $p(2) = R^0$ even if $(x,y) \in Ap(1)$. So 1 is not a dictator. And $t \neq 1$ is not a dictator: if $p \in L(X)^T$, $(x,y) \in Ap(1)$ and $(y,x) \in Ap(t)\ \forall\ t \neq 1$ then $(x,y) \in Af(p)$. The Pareto criterion is satisfied because we always have $f(p) = p(t)$ for some t. For the same reason $f(p)$ is always a preorder. To see that the independence axiom is satisfied suppose that $p(t)$ and $p'(t)$ agree over $\{a,b\}$ for all $t \in T$. If $p(t) \neq R^0$ for all t then $p'(t) \neq R^0$ for all t and hence $f(p) = p(1)$ and $f(p') = p'(1)$ so $f(p)$ and $f(p')$ agree over $\{a,b\}$. If $p(t) = R^0$ then $p(t) \cap \{a,b\}^2 = p'(t) \cap \{a,b\}^2$ implies $p'(t) = R^0$ and hence $f(p) = p(2)$ and $f(p') = p'(2)$. Again $f(p)$ and $f(p')$ agree over $\{a,b\}$. This is not a counter-example to Theorem 5 because D does not have the free-triple property. There is no $R \in D_t$ such that $(x,y) \in SR$ and $(x,z) \in AR$. Let us see why the proof of Theorem 5 does not apply to this example. Set $p(t) = R^0$ for all t and choose p' so that $Ap'(t) = \{(x,y),(x,z),(y,z)\}$ for all t. Both p and p' belong to D. If p'' agrees with p over $\{x,y\}$ and with p' over $\{y,z\}$ then p'' does not belong to D, although $P(X)^T$ contains a suitable p''. We have $(x,y) \in Sp''(t)$ but $p''(t) \neq R^0$.

The social welfare function of Example 7 is well defined over the domain $P(X)^T$ but it does *not* satisfy Arrow's independence axiom on that domain. Define p by setting $p(1) = R^1$ for $AR^1 = \{(x,y), (y,z), (x,z)\}$ and $p(t) = R^2\ \forall\ t \neq 1$ with $AR^2 = \{(x,z), (y,z)\}$. Define p' by setting $p'(1) = R^1$ and $p'(t) = R^0$ for all $t \neq 1$. Then $f(p) = p(1) = R^1$ and $f(p') = p(2) = R^0$. We have (x,y)

$\in Af(p)$ but $(y,x) \in f(p')$ although $p(t) \cap \{x,y\}^2 = p'(t) \cap \{x,y\}^2$ for all $t \in T$.

We have just shown that enlargement of a domain can vitiate the Arrow and Wilson impossibility theorems. The example serves as a lesson in logic and a caveat to scholars, but it is not otherwise useful. Individual 1 has far too much power although he is technically not a dictator. Now we turn to a practically important example to show how domain *contraction* can vitiate the impossibility theorems.

8. Majority rule. D is any domain. Define $f: D \to B(X)$ by setting $(x,y) \in f(p)$ if and only if the set $\{t \in T \mid (x,y) \in p(t)\}$ has at least as many members as the set $\{t \in T \mid (y,x) \in p(t)\}$.

This social welfare function obviously satisfies Arrow's independence axiom and the Pareto criterion on any domain. It is non-dictatorial if for each $t \in T$ there exist $x,y \in X$ and $p \in D$ such that $(x,y) \in Ap(t)$ but $(y,x) \in Ap(i)$ $\forall \ i \neq t$ (and T has at least two members). However, $f(p)$ is not even acyclic for all profiles in $L(X)^T$. Suppose that $T = \{1,2,3\}$, $X = \{x,y,z\}$ and the profile p is represented by Table 1. That is, $p_i(t) = R^t$ for all t, and $(x,y),(y,z) \in AR^1$, etc. Then $(x,y) \in Af(p)$, $(y,z) \in Af(p)$, *and* $(z,x) \in Af(p)$. This is the famous paradox of voting discovered by Condorcet (1785) and so strikingly generalized by Arrow (1951). There are non-trivial domains over which majority rule generates preorders. For example $D_t = \{R^1, R^2, R^4, R^5\}$, $\forall \ t \in T$, with R^4 and R^5 specified by Table 2. If $T = \{1,2,3\}$ then $f(p)$ is a linear

TABLE 1

R^1	R^2	R^3
x	y	z
y	z	x
z	x	y

TABLE 2

R^4	R^5
y	z
x	y
z	x

order for each $p \in D$: because the number of individuals is odd and no individual is ever indifferent between distinct alternatives we can have $(a,b) \in Sf(p)$ only if $a = b$. Therefore, $f(p)$ is a linear order if it is transitive, and it will be transitive if and only if there is a majority winner. Consider any $p \in D$. If $p(t) \in \{R^1, R^5\}$ for at most one individual then at least two individuals will rank y at the top so y defeats both x and z by a majority. If $p(i) = R^1$ and $p(j) = R^5$ then these two orderings cancel each other and $f(p)$ will be identical to the third individual's ordering. If two persons have R^1 or two persons have R^5 there is again a clear majority winner.

The domain $\{R^1, R^2, R^4, R^5\}^{\mathrm{T}}$ is an example of single-peaked preferences. Black (1948) was the first to identify this family of preferences and to show that there always exists a clear majority winner within the family. Complete characterizations of the domains over which majority rule is acyclic (etc.) were developed by Sen and Pattanaik (1969) and others. See Sen (1986*b*) for full references.

We have seen that some free-triple assumption is essential to the impossibility theorems. The assumption that X has at least three alternatives is vital: majority rule will be fully transitive over $P(X)^{\mathrm{T}}$ if X has only two members, and it always satisfies non-dictatorship, the Pareto criterion, and the independence axiom. Arrow's independence axiom is also critical. If it is eliminated from the list of requirements and X is any finite set simply associate with each $R \in P(X)$ a specific real-valued function $U(R,\cdot)$ on X such that $U(R,x) \geq U(R,y)$ if and only if xRy. For any $p \in P(X)^{\mathrm{T}}$ define f by setting $(x,y) \in f(p)$ if and only if $\Sigma_{t \in T} U(p(t),x) \geq \Sigma_{t \in T} U(p(t),y)$. Then $f(p)$ is transitive: $\Sigma_{t \in T} U(p(t),x) \geq \Sigma_{t \in T} U(p(t),y) \geq \Sigma_{t \in T} U(p(t),z)$ implies $\Sigma_{t \in T} U(p(t),x) \geq \Sigma_{t \in T} U(p(t),z)$. The Pareto criterion is satisfied because if $(x,y) \in \cap_{t \in T} Ap(t)$ then $U(p(t),x) > U(p(t),y)$ for all $t \in T$ and thus $\Sigma_{t \in T} U(p(t),x) > \Sigma_{t \in T} U(p(t),y)$. And f is non-dictatorial. We prove this for the case $T = \{1,2\}$. Choose any $R \in L(X)$ and set $R' = -R$. Define p and p' by setting $p(1) = R = p'(2)$ and $p(2) = R' = p'(1)$. Suppose $(x,y) \in Ap(1)$ and $(x,y) \in Af(p)$. Then $U(R,x) + U(R',x) > U(R,y) + U(R',y)$. This implies that $(x,y) \in Af(p')$. Because $(y,x) \in Ap'(1)$ individual 1 is not a dictator. Because $(y,x) \in Ap(2)$ and $(x,y) \in Af(p)$ person 2 is not a dictator either. Therefore, f must violate Arrow's independence axiom (by Theorem 5). To illustrate, we define a specific function U.

9. Borda's rule. X is finite. For each $R \in P(X)$ and $x \in X$ let $U(R,x)$ be the negative of the number of alternatives $y \in X$ such that $(y,x) \in AR$. Set $(x,y) \in f(p)$ if and only if $\Sigma_{t \in T} U(p(t),x) \geq \Sigma_{t \in T} U(p(t),y)$.

To illustrate a violation of the independence axiom take $T = \{1,2\}$ and $X = \{x,y,z\}$. Consider the profiles p and p' defined by Table 3. $U(p(1),x) + U(p(2),x) = -2 = U(p(1),y) + U(p(2),y) = U(p'(1),y) + U(p'(2),y)$, but $U(p'(1),x) + U(p'(2),x) = -3$. Therefore $(x,y) \in f(p)$ and $(y,x) \in Af(p')$, although $(y,x) \in Ap(1) \cap Ap'(1)$ and $(x,y) \in Ap(2) \cap Ap'(2)$. (Borda's rule is more fully discussed in Black (1958), and a thorough discussion of Borda in relation to Arrow's independence axiom may be found in Bordes and Tideman (1991).)

TABLE 3

$p(1)$	$p(2)$	$p'(1)$	$p'(2)$
z	x	y	z
y	y	z	x
x	z	x	y

We conclude this chapter with a lengthy comment on the infinite society case. From the standpoint of social choice theory the most important example of an infinite society economy is the overlapping generations model of Samuelson (1958). There are an infinite number of generations each of which lives for a finite number of periods, overlapping the lifetimes of other generations. Because T is infinite it cannot be assumed that the intersection of all decisive coalitions is decisive and the proofs of Theorems 4 and 5 break down, although Lemmas 1–3 and 4.3–4.5 remain valid because they do not depend on the finiteness assumption.

The discussion centres on the structure of decisive coalitions. We begin by defining filters and ultrafilters.

10. Filter. A collection F of subsets of T is a filter on T if:

10a $\quad \emptyset \notin F$ and $T \in F$.

10b $I \in F$ if $J \in F$ and $J \subset I$.
10c $I \cap J \in F$ if $I \in F$ and $J \in F$.

Note that 10a and 10c imply that a set and its complement cannot both belong to F. Given a social welfare function f let U_f denote the family of coalitions that are decisive for f. If U_f is non-empty then it satisfies 10a and 10b by definition of decisiveness, and U_f satisfies 10c if f satisfies the hypothesis of Lemmas 4.4–4.5. In particular, full transitivity of social preference is not required for the demonstrations that U_f has property 10c: quasitransitivity suffices. Similarly, V_f, the family of inversely decisive coalitions, satisfies 10a, 10b, and 10c if f satisfies the hypothesis of Lemmas 4.4–4.5 and V_f is not empty. If social preference is fully transitive then U_f and V_f have a finer structure, as we now show.

11. Ultrafilter. A collection F of subsets of T is an ultrafilter if it is a filter and

11a $I \in F$ if $I \subset T$ and $T - I \notin F$.

We now have all the ingredients for the following basic result.

12. Lemma. Let T be any set. Assume that X has at least three alternatives and D has the free-triple property. If $f: D \to P(X)$ is a non-null social welfare function satisfying non-imposition and Arrow's independence axiom then either U_f or V_f is an ultrafilter.

Proofs. Either U_f or V_f is non-empty by Lemma 1. Lemmas 4.3–4.5 establish 10a, 10b, and 10c for whichever of U_f or V_f is not empty. Lemmas 2 and 3 establish property 11a. ∎

An ultrafilter F is *free* if $\cap F = \varnothing$ and it is *fixed* if $\cap F \neq \varnothing$. Suppose that $t^* \in T$ and $t^* \in \cap F$. If $T - \{t^*\}$ belongs to F then $\cap F = \cap F \cap (T - \{t^*\})$, contradicting $t^* \in \cap F$. Therefore, $t^* \in \cap F$ implies that $T - \{t^*\}$ does not belong to F, and hence $\{t^*\}$ belongs to F by property 11a. Therefore, $\cap F = \{t^*\}$ if $t^* \in \cap F$. Either an ultrafilter is free or there is some $t^* \in T$ such that $F = \{I \in T \mid t^* \in I\}$. This means that f is dictatorial if U_f is a fixed ultrafilter and f is inversely dictatorial if V_f is a fixed ultrafilter. Fishburn (1970) shows how a free ultrafilter on an infinite T can be used to define a non-dictatorial social welfare function $f: P(X)^T \to P(X)$ satisfying the Pareto criterion and Arrow's independence axiom. (An ultrafilter on a finite set cannot be free as a consequence of 10a and 10c.)

13. Proposition. Assume that X has the discrete topology and U is a free ultrafilter on T. Then there is a nonauthoritarian social welfare function $f: P(X)^T \to P(X)$ satisfying the Pareto criterion, Arrow's independence axiom, and $U_f = U$.

Proof. Set $(x,y) \in f(p)$ if and only if the set $\{t \in T \mid (y,x) \in Ap(t)\}$ does not belong to U. Because $\varnothing \notin U$ and $T \in F$ the social welfare function satisfies the Pareto criterion, and it satisfies the independence axiom by definition. If $I \in U$ and $(x,y) \in \cap_I Ap(i)$ then $J \equiv \{t \in T \mid (y,x) \in Ap(t)\} \subset T - I$. Because $T - I$ does not belong to U the set J does not belong either by 10b. Therefore $(x,y) \in Af(p)$ by definition and hence $I \in U_f$. That is, $U \subset U_f$. If $I \in U_f$ but $I \notin U$ then $T - I \in U$ and thus $T - I \in U_f$. By definition, a set and its complement cannot both belong to U_f. Therefore, $U_f \subset U$. The social welfare function f is not inversely dictatorial because it satisfies the Pareto criterion. It is not dictatorial because $\cap U_f = \varnothing$ implies that $\{t\}$ does not belong to U_f for any $t \in T$.

It remains to show that $f(p)$ is a preorder. If $(x,y) \notin f(p)$ then $I = \{t \in T \mid (y,x) \in Ap(t)\} \in U$ and hence $T - I \notin U$. $J = \{t \in T \mid (x,y) \in Ap(t)\} \subset T - I$ and thus $J \notin U$ by10b. Therefore $(y,x) \in f(p)$ and $f(p)$ is complete. Suppose that (x,y) and (y,z) both belong to $f(p)$. Set $I = \{t \in T \mid (x,y) \in p(t)\}$ and $J = \{t \in T \mid (y,z) \in p(t)\}$. Then I and J belong to U by 11a and the definition of f, and thus $I \cap J \in U$ by property 10c. By transitivity of $p(t)$ we have $(x,z) \in p(t)$ for all $t \in I \cap J$. Because $T - (I \cap J) \notin U$, no subset of $T - (I \cap J)$ belongs to U by property 10b. Therefore, the set $\{t \in T \mid (z,x) \in Ap(t)\}$ does not belong to U and therefore $(x,z) \in f(p)$ by definition. This proves that $f(p)$ is transitive. ∎

Kelly (1978: 112) uses Zorn's lemma to prove that there is a free ultrafilter associated with every infinite set. Therefore, Proposition 13 is a *possibility* theorem for infinite societies, at least when there is no meaningful continuity restriction on social preference. Chapter 9 will show that free ultrafilters are incompatible with continuity of social preference in a sensible topology.

6 Quasitransitive Social Preference

IN 1970 A.K. Sen pointed out that the hypothesis of Arrow's impossibility theorem admits a non-dictatorial social welfare function if the requirement of full transitivity of social preference is weakened to quasitransitivity (Sen 1970). Simply set x above y in the social preference scheme if and only if every individual strictly prefers x to y. We now define this procedure formally.

1. Pareto aggregation rule. For any $p \in D$ and any $x,y \in X$ set $(x,y) \in f(p)$ if and only if $(x,y) \in p(t)$ for some $t \in T$.

The grand coalition T is obviously an oligarchy for the Pareto aggregation rule.

2. Proposition. Let T be any set of individuals and let X be any set of outcomes. The Pareto aggregation rule satisfies the Pareto criterion and Arrow's independence axiom, and $f(p)$ is quasitransitive for any profile p of quasitransitive individual preference relations. The rule is non-dictatorial if for each $i \in T$ there is some $j \in T$, some $p \in D$, and some $x,y \in X$ such that $(x,y) \in Ap(i) \cap -Ap(j)$.

Proof. Obviously, $f(p) = \cup_{t \in T} p(t)$ and $Af(p) = \cap_{t \in T} Ap(t)$ for any profile p. The former implies that $f(p)$ is complete and the latter implies that f satisfies the Pareto criterion. Arrow's independence axiom is obviously satisfied by this rule. If $(x,y) \in Af(p)$ and $(y,z) \in Af(p)$ then (x,y) and (y,z) both belong to $\cap_{t \in T} Ap(t)$. Therefore, (x,z) belongs to $\cap_{t \in T} Ap(t)$ if each $p(t)$ is quasitransitive, and hence $(x,z) \in Af(p)$.

If $(x,y) \in Ap(i)$ and $(y,x) \in Ap(j)$ then $(x,y) \in f(p) \cap -f(p) = Sf(p)$, and neither i nor j can be a dictator. ∎

Gibbard (1969), Mas-Colell and Sonnenschein (1972), and Guha (1972) independently proved that properly oligarchical functions are the *only* admissible social welfare functions when Arrow's hypothesis is relaxed slightly to allow intransitive social indifference. We begin with a generalization of this result by weakening the Pareto criterion to strict non-imposition. Part of our preliminary result was first discovered by Fountain and Suzumura (1982). As in Chapter 5, any topology on X may be assumed as long as it is consistent with the free-triple property.

3. Theorem. Assume that T is finite, X has at least three alternatives, and D has the free-triple property. If $f: D \to Q(X)$ satisfies strict non-imposition and Arrow's independence axiom then f is quasi-oligarchical.

Proof. Lemma 4.13 provides the proof (see Definition 3.24.) ∎

Example 4.6 reveals that a quasi-oligarchical social welfare function need not be oligarchical. In this example $(\{1\}, \varnothing, \{2\})$ is a decisive triple but $(\{1\}, \varnothing)$ is not a decisive pair. If *individual* indifference can be intransitive then we actually have an oligarchy, as we now prove.

4. Theorem. Assume that T is finite and X has at least three alternatives. If $f: Q(X)^{\mathrm{T}} \to Q(X)$ satisfies strict non-imposition and Arrow's independence axiom then f is oligarchical.

Proof. By Theorem 3, f is at least quasi-oligarchical. Let (I, J, K) be the quasi-oligarchy for f. Suppose $K \neq \varnothing$. Choose three distinct alternatives $x, y, z \in X$ and $p \in D$ arbitrarily except that (x, y), $(y, z) \in Ap(t)$ for all $t \in I$, (z, y), $(y, x) \in Ap(t)$ for all $t \in J$, (x, y), $(y, z) \in Sp(t)$ for all $t \in K$, and $(z, x) \in Ap(t)$ for all $t \in K$. Then (x, y) and (y, z) belong to $Af(p)$ because (I, J, K) is decisive, but $(x, z) \notin Af(p)$ because the members of K have quasi-veto power. Therefore, $f(p)$ is not quasitransitive. ∎

From now on we maintain the assumption that each individual preference relation is a preorder. *If* there exists a decisive pair of coalitions then the social welfare function is oligarchical, but a decisive pair may not exist (Example 4.6).

5. Theorem. Assume that T is finite, X has at least three alternatives, and D has the free-triple property. If $f: D \to Q(X)$ satisfies strict non-imposition and Arrow's independence axiom and there exists a decisive pair of coalitions then f is oligarchical.

Proof. Because T is finite and there exists a decisive pair of coalitions there exists a minimal decisive pair (I^*, J^*) by Lemma 4.5.

Each member of I^* has veto power and each member of J^* has inverse veto power by Lemma 4.8. ∎

There are two assumptions that will be used to guarantee the existence of a decisive pair of coalitions. Continuity of social preference is employed in Chapters 8 – 13. Instead one can impose the Pareto criterion as in the next theorem, which was first proved by Gibbard (1969), Mas-Colell and Sonnenschein (1972), and Guha (1972).

6. Theorem. Assume that T is finite, X has at least three alternatives, and D has the free-triple property. If $f: D \to Q(X)$ satisfies the Pareto criterion and Arrow's independence axiom then f is properly oligarchical.

Proof. The pair $(T, \varnothing)$ is decisive by the Pareto criterion. Then there is some minimal decisive pair (I^*, J^*) for f by Lemma 4.5, and (I^*, J^*) is an oligarchy for f by Lemma 4.8. By definition of a minimal decisive pair $(I^*, J^*) = (I^* \cap T, J^* \cap \varnothing)$. Therefore, $J^* = \varnothing$, and thus f is properly oligarchical. ∎

There is no need to consider infinite society counter-examples to these impossibility theorems. Proposition 5.13 points to the existence of a nonauthoritarian social welfare function with an unrestricted domain and satisfying the Pareto criterion and Arrow's independence axiom *and* yielding fully transitive social preferences. Quasitransitivity of social preference is automatically satisfied. (Show that the social welfare function of 5.13 cannot be oligarchical.)

7 The Gibbard–Satterthwaite Theorem

THIS chapter concerns the properties of voting schemes that induce individuals to report their preferences truthfully. A voting scheme selects a single alternative for a given profile of reported individual preferences. The selected alternative is the winner of the election represented by the voting scheme. Consider a classic example, the Borda scheme.

1. Borda's voting scheme. There are m alternatives. Alternative x receives $V(R,x)$ votes from individual t if there are exactly $m - V(R,x)$ alternatives ranking above x in t's reported preference scheme R. If p is the profile of reported preferences then alternative x's score is $\Sigma_{t \in T} V(p(t),x)$ and the winner is the alternative with the highest score. (Ties are broken in some non-random fashion.)

To illustrate, suppose that all preferences are linear. If alternative x is t's most-preferred alternative then x will receive m votes from t. The second-most-preferred alternative will receive $m - 1$ votes from t, and so on. Suppose there are four alternatives w,x,y,z, and three individuals 1,2,3. The true and reported preferences are given by Table 4. (Alternatives are listed under a person's name in decreasing order of preference: person 1 prefers w to x, x to y, and y to z according to her true preferences. The other columns are interpreted similarly.)

TABLE 4

True preferences			Reported preferences		
1	2	3	1	2	3
w	z	y	w	z	x
x	w	x	x	w	y
y	x	w	y	x	z
z	y	z	z	y	w

If everyone reports, i.e. votes according to, her true preferences the winner will be w with 9 votes: 4 votes from person 1, 3 votes from 2, and 2 votes from 3. (Alternative x receives 8 votes, with 7 for y and 6 for z.) It is not in everyone's interest to report truthfully in every situation. For example, if 1 and 2 report truthfully then 3 can profit by reporting the preference scheme assigned to her in the last column, with x top-ranked. With these reported preferences alternative x will win with 9 votes (8 for w, 7 for z, and 6 for y.) Person 3 prefers x to w *according to her true preferences*, and has profited from the misrepresentation. Gibbard (1973) and Satterthwaite (1975) independently proved that *no* voting scheme is invulnerable to this sort of manipulation unless it is dictatorial or there are less than three eligible alternatives! Before proving this we need some new definitions and notation. (The reader is referred to Kelly (1978 ch. 6) for a brief history of the theory of manipulation. Kelly (1987 chs. 10 and 11) provides substantial motivation for the treatment of strategic voting.)

Assume a set X of alternatives. Continuity of preference will not play a role in this chapter so we suppose that X is endowed with the discrete topology. T is the set of individuals. For most of the results T will be finite. In fact we can prove the Gibbard–Satterthwaite theorem quite easily by assuming that there are only two individuals and then extending the result to the general case by means of a separate argument. Recall that $P(X)$ is the set of all preorders on X — we are assuming the discrete topology. As usual a domain D is a subset of $P(X)^T$ and we will consider only product sets of the form $D = \Pi_{t \in T} D_t$. Recall that $L(X)$ is the family of linear orders on X (1.24). We will assume that D_t is a subset of $L(X)$ for simplicity. The reader who wants the fully general result with $L(X)^T \subset D \subset P(X)^T$ and an arbitrary finite T can stay the course. Otherwise one can get the gist by stopping at Theorem 14 which assumes $T = \{1,2\}$ and $D = L(X)^T$.

2. Voting scheme. A voting scheme is a function $g: D \to X$. The outcome function g selects an alternative $g(p)$ in X for each profile p in D. Let X_g denote $\{g(p) \mid p \in D\}$, the *image* of g.

We should really think of the pair (D,g) as the voting scheme but there will be no confusion when we identify the outcome function with the voting scheme.

If R is a binary relation on X and $Y \subset X$ then $\max R(Y) =$

$\{x \in Y \mid (y,x) \notin AR$ for any $y \in Y\}$ (1.11). That is, max$R(Y)$ is the set of maximal elements in Y according to R. To streamline the exposition, in this chapter we will let max$R$ represent the set max$R(X)$, and $C(R,Y) = \{x \in Y \mid (y,x) \notin AR$ for any $y \in Y\}$ will represent max$R(Y)$ if Y is a subset of X. If $C(R,Y)$ or max$R$ is a singleton we will often identify the set with the element it contains — writing max$R = x$, etc.

3. Dictatorship. The voting scheme $g: D \to X$ is dictatorial if there is some individual $t \in T$ such that $g(p) \in C(p(t),X_g)$ for all $p \in D$. In words, g is dictatorial if every outcome is best for t among all the alternatives that can be selected as outcomes. Of course, individual t is called a dictator in that case.

4. Manipulation. The voting scheme $g: D \to X$ can be manipulated by individual t at profile p via the relation R if $p \in D$, $R \in D_t$ and $(g(p'),g(p)) \in Ap(t)$ for p' defined by setting $p'(t) = R$ and $p'(i) = p(i) \; \forall i \neq t$. Accordingly, the voting scheme is non-manipulable if for all $p \in D$ and $t \in T$ we have $(g(p),g(p')) \in p(t)$ for all $p' \in D$ such that $p'(i) = p(i)$ for all $i \neq t$.

In words, a voting scheme can be manipulated if in some situation there is some individual who by reporting a preference scheme other than her true preference relation can precipitate the selection of an outcome that is preferred *according to her true ordering* to the one that emerges when she announces her true ordering. In definition 4, $p'(i)$ plays the role of i's reported preference scheme and $p(i)$ is interpreted as i's true preference relation.

A voting scheme can be generalized to allow the transmission of messages that are not preference relations on X, although an individual's choice of message will still be governed by her true preference ordering. The reader who is only interested in the basic result on voting schemes can skip the following discussion of mechanisms and pick up the argument at Lemma 8.

If we suppose that a social welfare function $f: D \to B(X)$ has already been agreed to and X itself is the agenda of feasible alternatives then the socially best alternatives are the numbers of max$f(p)$, where p represents the profile of actual individual preferences for the social decision problem at hand. The next step is to discover what those preferences are and thus to identify a member of max$f(p)$. At this point a voting procedure is required.

We want a scheme (i) for which individuals can be relied on to report truthfully and (ii) which satisfies $g(p) \in \max f(p)$ for all $p \in D$. We do not impose the second requirement in this chapter because we are going to prove that the first cannot be satisfied unless g is dictatorial or there are only two available alternatives. Nevertheless, the motivation is important: we want to implement f. We may, however, have more success with a *mechanism* in which individuals are asked to announce generalized messages which may not be preference relations.

5. Mechanism. A mechanism is a pair (S,g), where S specifies a strategy set S_t for each $t \in T$ and $g: S \to X$ is an outcome function. If $s \in S$ is the profile of announced strategies then individual t has selected strategy $s(t) \in S_t$, and the outcome $g(s) \in X$ is realized. S is often implicit, in which case we identify g with the mechanism.

There are two reasons why we might not want $S_t = D_t$. First, it may be more economical to have individuals announce some summary statistic rather than an entire preference relation, just as consumers in a market economy 'announce' a vector of marginal rates of substitution rather than an entire utility function. Secondly, we may have more success implementing non-dictatorial social welfare functions if we allow more general strategy sets. We will see, however, that only the dictatorial mechanisms are successful. The generalized notion of non-manipulability rests on the idea of a *dominant strategy*.

6. Dominant strategy. If (S,g) is a mechanism and $R \in P(X)$ then $s^*(t) \in S_t$ is a dominant strategy for individual t with preference scheme R if $g(s)Rg(s')$ for all $s, s' \in S$ such that $s(t) = s^*(t)$ and $s'(i) = s(i)$ for all $i \neq t$.

Simply put, t has a dominant strategy $s^*(t)$ with respect to preference scheme R if the announcement $s^*(t)$ by t produces the outcome ranked highest by R *given* the messages of the other individuals. A dominant strategy *mechanism* is one for which each individual has a dominant strategy for each admissible preference relation.

7. Dominant strategy mechanism. The mechanism (S,g) is a dominant strategy mechanism with respect to $D \subset P(X)^{\mathrm{T}}$ if for each t

and each $R \in D_t$ there is a dominant strategy $\sigma_t(R)$ in S_t. The function σ_t is called a behaviour rule.

If the message space S and domain D are implicit we will simply refer to a dominant strategy mechanism g. By designing a mechanism with dominant strategies one can be sure that individuals will play according to the rules. The rule governing individual t's behaviour is: announce $\sigma_t(R)$ if your preference ordering is R. The rule also ensures a speedy convergence to the socially optimal outcome: t's best strategy, given her preference scheme, is independent of what the others announce so there will be no revision of messages after the first round of play. As the Gibbard–Satterthwaite theorem confirms, dominant strategies are extremely demanding.

One more definition is required. For $T = \{1,2\}$ define the *option correspondence* O_1 of individual 1 associated with voting system g. For each $R \in D_2$, $O_1(R) = \{g(p) \mid p \in D \text{ and } p(2) = R\}$, the set of alternatives that 1 can precipitate when 2 announces preference relation R.

We begin with a theorem on voting schemes from which the impossibility theorem for mechanisms will emerge as a simple corollary. Six preliminary lemmas are required; Proposition 1.37 is used frequently, but without acknowledgement. Each lemma assumes a domain D that is a product set and a society $T = \{1,2\}$ consisting of two individuals. The proof is due to Barberà (1983*a*). (Barberà (1983*b*) offers a proof of Arrow's theorem using a similar technique.)

8. Lemma. Suppose that $D \subset L(X)^{\{1,2\}}$ and $g: D \to X$ is a non-manipulable voting scheme. If $p \in D$ then there is no $y \in O_1(p(2))$ such that $(y, g(p)) \in Ap(1)$.

Proof. Suppose that $g(p) = x$ and y belongs to $O_1(p(2))$. Then there is some $p' \in D$ such that $p'(2) = p(2)$ and $g(p') = y$. If $(y, x) \in Ap(1)$ then 1 can manipulate g at p via $p'(1)$. Therefore, there is no $y \in O_1(p(2))$ such that $(y, g(p)) \in Ap(1)$. ∎

9. Lemma. Suppose that $D \subset L(X)^{\{1,2\}}$ and $g: D \to X$ is a non-manipulable voting scheme. Then for all $R \in D_2$ we have $C(R, X_g) \subset O_1(R)$.

Proof. Suppose $C(R, X_g) = \{x\}$. Then there is some $p \in D$ such that $g(p) = x$. If $g(p') = y$ for $p'(1) = p(1)$ and $p'(2) = R$ and $x \neq y$ then $(x, y) \in AR$. Therefore, 2 can manipulate g at p'

via $p(2)$, a contradiction. Therefore, $g(p') = x$ and hence $x \in O_1(R)$. ∎

10. Lemma. Suppose that $D \subset L(X)^{\{1,2\}}$ and $g: D \to X$ is a non-manipulable voting scheme. For all $p \in D$, we have $g(p) = C(p(1), X_g)$ whenever $C(p(1), X_g) = C(p(2), X_g)$.

Proof. Suppose that $C(p(t), X_g) = \{x\}$ for all $t \in \{1,2\}$. Then $x \in O_1(p(2))$ by Lemma 9 and therefore $g(p) = x$ by Lemma 8. ∎

11. Lemma. Suppose that X has at least three members and $g: L(X)^{\{1,2\}} \to X$ is non-manipulable. Then $O_1(R) = O_1(R')$ for all $R, R' \in D_2$ such that $C(R, X_g) = C(R', X_g)$.

Proof. Suppose that $C(R, X_g) = C(R'X_g) = \{x\}$. Suppose $y \in O_1(R)$ and $y \notin O_1(R')$. By Lemma 9 we have $x \in O_1(R)$ $\cap O_1(R')$. Let $p \in D$ satisfy $p(2) = R$ and (y,x), $(x,z) \in p(1)$ for all $z \in X$ such that $z \notin \{x,y\}$. Let $p' \in D$ satisfy $p'(1) = p(1)$ and $p'(2) = R'$. By Lemma 8 we have $g(p) = y$ and $g(p') = x$. But then individual 2 can manipulate g at p via $p'(2)$, a contradiction. Therefore, $O_1(R) = O_1(R')$. ∎

12. Lemma. Assume that X_g has at least three members and $g: L(X)^{\{1,2\}} \to X$ is a non-manipulable voting scheme. For all $R \in L(X)$ either $O_1(R) = X_g$ or $O_1(R)$ is a singleton.

Proof. Suppose there exist $R \in L(X)$ and distinct $x,y,z \in X_g$ such that $x,y \in O_1(R)$ but $z \notin O_1(R)$. Then $z \notin C(R, X_g)$ by Lemma 9. Choose $R' \in L(X)$ such that $\max R' = C(R, X_g)$ and $(z,w) \in R'$ for all $w \in X - C(R, X_g)$. Then $O_1(R') = O_1(R)$ by Lemma 11. Therefore, $z \notin O_1(R')$ and $x,y \in O_1(R')$. Without loss of generality we can assume that $\max R' = x$ (Lemma 9). Then we have $(x,z) \in AR'$ and $(z,y) \in AR'$. Choose $R'' \in L(X)$ such that $\max R'' = z$ and $(y,w) \in AR''$ if $w \notin \{z,y\}$. Set $p(1) = R''$ and $p(2) = R'$. Then $g(p) = y$ by Lemma 8. Now choose $p' \in L(X)^{\mathrm{T}}$ such that $\max p'(2) = z$ and $p'(1) = R''$. Then $g(p') = z$ by Lemma 10. Then person 2 can manipulate g at p via $p'(2)$. The contradiction establishes the lemma: if z belongs to X_g but not to $O_1(R)$ then $O_1(R)$ cannot contain two distinct alternatives. ∎

13. Lemma. Assume that $g: L(X)^{\{1,2\}} \to X$ is non-manipulable and X_g has at least three members. Either $O_1(R)$ is a singleton for all $R \in L(X)$ or else $O_1(R) = X_g$ for all $R \in L(X)$.

Proof. Suppose that $O_1(R) = X_g$ and $O_1(R') = \{x\}$ for R, $R' \in L(X)$ and $x \in X$. Choose $R'' \in L(X)$ as follows: $R'' = R$ if $\max R = x$, and if $\max R = y \neq x$ choose any $R'' \in L(X)$ such that $\max R'' = y$ and $(x,z) \in R''$ for all $z \in X - \{y\}$. Then $O_1(R'') = O_1(R)$ by Lemma 11. Beacause X_g has at least three members and $O_1(R'') = X_g$ we have $g(p) = z$ for some $p \in D$ such that $p(2) = R''$ and some $z \in X$ such that $(x,z) \in AR''$. But $O_1(R') = \{x\}$ so $g(p') = x$ if $p'(1) = p(1)$ and $p'(2) = R'$. Then person 2 can manipulate g at p via R'. Because g is non-manipulable and Lemma 12 applies we must have $O_1(R) = X_g$ for *all* $R \in D_2$ or else $O_1(R)$ is a singleton for all $R \in D_2$. ∎

14. Theorem. Assume that $g: L(X)^{[1,2]} \to X$ is non-manipulable. Then g is dictatorial if X_g has at least three members.

Proof. Suppose that $O_1(R)$ is a singleton for all $R \in L(X)$. Then $O_1(R) = C(R, X_g)$ for all $R \in L(X)$ by Lemma 9. Then $g(p) = C(p(2), X_g)$ for all $p \in L(X)^{[1,2]}$ and person 2 is a dictator. If person 2 is not a dictator then we have $O_1(R) = X_g$ for all $R \in L(X)$ by Lemma 13. But then person 1 is a dictator because $g(p) = C(p(1), X_g)$ for all $p \in L(X)^{[1,2]}$ by Lemma 8. ∎

We will show that the impossibility theorem actually holds for any finite set T and any product domain D containing $L(X)^T$. Then we go on to prove that all dominant strategy mechanisms are dictatorial. We begin by showing that for *any* finite set T a non-manipulable voting scheme $g: L(X)^T \to X$ is dictatorial if X_g has at least three members. Our technique for extending Theorem 14 to arbitrary finite T is based on a proof in Kalai and Muller (1977), although our argument is more direct. (The Kalai and Muller paper is more ambitious. It characterizes the domains on which non-manipulability implies dictatorship.)

15. Lemma. Let $T = \{1, 2, \ldots, n\}$. If $g: L(X)^T \to X$ is non-manipulable then for all $x \in X_g$ and all $p \in L(X)^T$, $g(p) = x$ if $x = \max p(t)$ for all $t \in T$.

Proof. Suppose $x \in X_g$ and $\max p(t) = x$ for all $t \in T$, but $g(p) = y \neq x$. We have $g(p') = x$ for some $p' \in L(X)^T$. Define the sequence $p^0, p^1, \ldots, p^n$ in $L(X)^T$ inductively. Set $p^0 = p'$, $p^i(i) = p(i)$, and $p^i(t) = p^{i-1}(t)$ for $t \neq i$ and $i > 0$. Then $g(p^0) = x \neq y = g(p^n)$. Let j be the largest integer such that $g(p^j) = x$. Then individual $j + 1$ can manipulate g at p^{j+1} via $p^j(j + 1)$, a contradiction. Therefore, $g(p) = x$. ∎

16. Theorem. Assume that $T = \{1,2,\ldots,n\}$ and that g: $L(X)^{\mathrm{T}} \to X$ is non-manipulable. Then g is dictatorial if X_g has at least three members.

Proof. If T is a singleton then g is obviously dictatorial if it is non-manipulable. If $n = 2$ and g is non-manipulable then g is dictatorial by Theorem 14. Assume that the theorem is true for the n-person case. We prove Theorem 16 by proving the result for $n + 1$ persons, with $n \geq 2$.

Set $T = \{1,2,\ldots n,n + 1\}$ and choose arbitrary $i \in T$. Set $N = T - \{i + 1\}$ if $i \leq n$ and $N = \{2,3,\ldots,n,n + 1\}$ if $i = n + 1$. Define g^i: $L(X)^{\mathrm{N}} \to X$ by setting $g^i(p) = g(p')$, where $p' \in L(X)^{\mathrm{T}}$ satisfies $p'(t) = p(t)$ if $t \neq i + 1$ and $p'(i + 1) = p(i)$. (Of course, $n + 2$ denotes individual 1.) We have $\{g^i(p) \mid p \in L(X)^{\mathrm{N}}\} = X_g$ by Lemma 15. Then by the induction hypothesis g^i is dictatorial if X_g has at least three members. Suppose that t is a dictator for g^i and $t \neq i$. If t is not a dictator for g then $g(p^1) = x \notin C(p^1(t),X_g)$ for some $p^1 \in L(X)^{\mathrm{T}}$. Define $p^2 \in L(X)^{\mathrm{T}}$ by setting $p^2(j) = p^1(j) \, \forall j \neq i$ and $p^2(i) = p^2(i + 1) = p^1(i + 1)$. Set $y = g(p^2)$. Then $C(p^1(t),X_g) = C(p^2(t),X_g) = \{y\}$ because t is a dictator for g^i and $i \neq t \neq i + 1$. Therefore $(x,y) \in Ap^1(i)$ because i cannot manipulate g at p^1 via $p^2(i)$. But $g(p^3) = y$ if $p^3(j) = p^1(j) \, \forall j \neq i + 1$ and $p^3(i + 1) = p^1(i)$, because t is a dictator for g^i. Then $i + 1$ can manipulate g at p^3 via $p^1(i + 1)$, a contradiction. Therefore, $t \neq i$ cannot be a dictator for g^i if g is non-manipulable. By the induction hypothesis person i must be a dictator for g^i.

We have proved the following: if g is non-manipulable and non-dictatorial then for all $i \in T$ and $p \in L(X)^{\mathrm{T}}$ we have $g(p) = C(p(i),X_g)$ if $p(i + 1) = p(i)$. This follows from the fact that i is a dictator for g^i if g is non-manipulable and non-dictatorial. If $n + 1 \geq 4$ choose distinct $x,y \in X_g$ and $p \in L(X)$ so that $p(1) = p(2)$, $p(n) = p(n + 1)$, $\max p(1) = x$, and $\max p(n) = y$. Then $g(p) \in C(p(1),X_g) \cap C(p(n),X_g)$ if 1 is a dictator for g^1 and n is a dictator for g^n. Obviously, this is impossible. Therefore, we have our result for all finite T if we can prove it for $T = \{1,2,3\}$.

Assume that $T = \{1,2,3\}$ and choose three distinct alternatives x^1, x^2, $x^3 \in X_g$ and $R^t \in L(X)$ such that (x^1,x^2), $(x^2,x^3) \in AR^1$, (x^2,x^3), $(x^3,x^1) \in AR^2$, and (x^3,x^1), $(x^1,x^2) \in AR^3$, and $(x^i,y) \in AR^t$ for all $i,t \in \{1,2,3\}$ and all $y \in X - \{x^1,x^2,x^3\}$. Define

p by setting $p(1) = R^1$, $p(2) = R^2$ and $p(3) = R^3$. We need to prove that $g(p)$ belongs to $Z = \{x^1, x^2, x^3\}$. Set $Y = X - Z$. Define profile q by setting $q(1) = q(2) = p(1)$ and $q(3) = p(3)$. Set $q'(t) = p(1)$ for all t. If $g(p) \in Y$ then $g(q) \in Y$, otherwise individual 2 can manipulate g at p via $q(2)$. And if $g(q)$ belongs to Y then so does $g(q')$; otherwise individual 3 can manipulate g at q via $q'(3)$. But $g(q') \in Y$ contradicts Lemma 15 because $\max q'(t) = x^1$ for all t. Therefore, $g(p) \in Z$. Set $g(p) = x^t$. Now, person $t + 1$ is a dictator for g^{t+1} so $g(p') = x^{t+2}$ for $p'(j) = p(j) \; \forall j \neq t + 1$, and $p'(t + 1) = p(t + 2)$. (If $t + i > 3$ then $t + i$ refers to individual $t + i - 3$.) But $(x^{t+2}, x^t) \in Ap(t + 1)$ and person $t + 1$ can manipulate g at p via $p'(t + 1)$, a contradiction. Therefore, non-manipulability of g forces us to drop the supposition that there is no dictator for g. ∎

Now we generalize Theorem 16 to an arbitrary product domain containing $L(X)^T$. This will allow individual indifference between distinct alternatives. The proof is based on one of the steps in Schmeidler and Sonnenschein (1978).

17. Theorem. If $T = \{1, 2, \ldots, n\}$, $L(X)^T \subset D$, $g: D \to X$ is non-manipulable, and X_g has at least three members then g is dictatorial.

Proof. Define the restriction $h: L(X)^T \to X$ of g to $L(X)^T$ by setting $h(p) = g(p)$ for all $p \in L(X)^T$. Then h is non-manipulable if g is. Therefore $X_h = X_g$ by Lemma 15. That is, h has the same image as g. Then h is dictatorial by Theorem 16. Let person 1 be the dictator for h. We show that 1 is also a dictator for g.

Suppose to the contrary that $p \in D$ and $g(p) \notin C(p(1), X_g)$. Choose any $p' \in L(X)^T$ such that $(x, y) \in Ap'(1) \cap -Ap'(t)$ for all $t \in T - \{1\}$ and all $x \in C(p(1), X_g)$ and all $y \in X$ not belonging to $C(p(1), X_g)$. (Proposition 1.37.) Then $h(p') \in C(p'(1), X_g) \subset C(p(1), X_g)$. Define the sequence $p^0, p^1, \ldots, p^n$ in D. Set $p^0 = p$ and $p^i(i) = p'(i)$ for all $i = 1, 2, \ldots, n$, and for $i > 0$ set $p^i(t) = p^{i-1}(t)$ for $t \neq i$. Then $p^n = p'$ and $g(p^n) = g(p')$. Let j be the smallest integer such that $h(p^j) \in C(p(1), X_g)$. If $j = 1$ then individual 1 can manipulate g at p via $p'(1)$. If $j > 1$ then individual j can manipulate g at p^j via $p(j)$. But g is non-manipulable, so person 1 must be a dictator for g. ∎

Our last result is an application of the *revelation principle* to

Theorem 17 to prove that every dominant strategy mechanism is dictatorial (See Dasgupta *et al.* (1979) or Repullo (1986) for a discussion of the relevation principle in social choice theory.) Theorem 17 proves that truthful preference revelation cannot be a dominant strategy in all situations if the voting scheme is non-dictatorial and its image contains at least three alternatives. Suppose that (S,g) is a dominant strategy mechanism with respect to domain D. Let $\sigma_t(R)$ be a dominant strategy for person t with preference scheme $R \in D_t$. This induces the following direct revelation mechanism g': $D \to X$. Simply set $g'(p) = g(\sigma p)$ where $s \equiv \sigma p$ is defined by setting $s(t) = \sigma_t(p(t))$. Clearly, g' is a voting scheme with domain D. If $(g'(p'),g'(p)) \in Ap(t)$ and $p'(i) = p(i) \ \forall i \neq t$ then $(g(s'),g(s)) \in Ap(t)$ for $s'(i) = s(i) = \sigma(p(i)) \ \forall i \neq t$, $s'(t) = \sigma_t(p'(t))$, and $s(t) = \sigma_t(p(t))$, contradicting the claim that $\sigma_t(p(t))$ is a dominant strategy for t and $p(t)$. Therefore, g' is a non-manipulable voting scheme. If there is no individual $t \in T$ such that $g(\sigma p) \in C(p(t),X_g) \ \forall p \in D$ then g' is also non-dictatorial, where X_g denotes the set $\{g(\sigma p) | p \in D\}$. Then we have proved the following.

18. Theorem. If $T = \{1,2, \ldots, n\}$, $L(X)^T \subset D$ and (S,g) is a dominant strategy mechanism with respect to domain D and behaviour rules $\sigma_1, \sigma_2, \ldots, \sigma_n$ then either there is some individual t such that $g(\sigma p) \in C(p(t),X_g) \ \forall p \in D$ or X_g contains exactly two alternatives.

We conclude this chapter by illustrating the significance of various assumptions by means of some simple examples. Each one is a voting scheme.

19. Example. Define $g: L(X)^T \to X$ by setting $g(p) = C(-p(1),X)$.

Clearly, person 1 controls the outcome, but only by reporting $-R$ when her preference is R can she do so to her advantage. Even though 1 has complete control the mechanism is manipulable according to Definition 4.

20. Example. $X = \{x_1,x_2,x_3,x_4,x_5\}$. For each $p \in L(X)^T$ set $g(p) = C(p(1), \{x_1,x_2,x_3\})$, person 1's most preferred alternative in the set $X_g = \{x_1,x_2,x_3\}$.

Obviously, $g(p) \neq \max p(1)$ for many $p \in D$. All our theorems are false if dictatorship is taken to mean $g(p) = \max p(1)$ for all p.

To disqualify schemes such as Example 20 we have broadened the definition of dictatorship. A voting scheme is dictatorial if 1 always gets her most-preferred alternative in X_g, the set of possible outcomes.

21. Example. $T = \{1,2,\ldots,n\}$, n is odd, and $X = \{x,y\}$. For $D = L(X)^T$ let $g(p)$ be the alternative that ranks first in the majority of preferences $p(t)$. Then g is non-manipulable and non-dictatorial.

We have just defined majority rule on a two-alternative set of outcomes. It is obviously not dictatorial. If $g(p) = x$ then t can change the outcome only if x is most preferred according to $p(t)$. But t would not want to change the outcome if $p(t)$ is her true preference scheme. Therefore, majority rule is non-manipulable if there are only two alternatives. The assumption that X_g has more than two members is crucial to the impossibility theorems, as is the assumption that D_t contains all the *linear* orders on X. To illustrate the last point, suppose that X has more than two alternatives but $D \subset L(X)^T$ is restricted so that there is always a single majority winner. That is, for each $p \in D$ there is some $x \in X$ such that the set $\{t \in T \mid (x,y) \in p(t)\}$ contains over half the members of T for each y in X. (Refer to Example 5.8 and the accompanying discussion.) Then the voting scheme that selects the majority winner is non-manipulable. If $g(p) = x \neq y = g(p')$ and $p'(i) = p(i)$ for all $i \neq t$ then x is the majority winner when t reports $p(t)$ and y is the majority winner when t reports $p'(t)$ and all other individual preference announcements are unchanged. If y defeats x when t reports $p'(t)$ but x defeats y when t reports $p(t)$ truthfully then t must cast a vote for x over y under truthful revelation: t truly prefers x to y.

Suppose, however, that $T = \{1,2,3\}$, $X = \{x,y,z\}$, and $g(p) = z$, which we take to be the status quo, if there is no clear majority winner. Let the domain be defined by $D_t = \{R^1, R^2, R^4, R^5\}$ for $t = 1,2,3$, where R^t is specified in Tables 1 and 2 of Chapter 5. Then there is a clear majority winner for each $p \in D$. As we have just argued (D,g) is non-dictatorial and non-manipulable (according to Definition 4) for the majority-rule voting scheme g. Define $p \in D$ by setting $p(1) = R^1$, $p(2) = R^2$, and $p(3) = R^5$. Then $g(p) = y$. Although R^3 does not belong to D_3, if 3 announces R^3 the outcome will be z as long as 1 and 2 continue to announce R^1

and R^2 respectively, and 3 prefers z to y according her true preference scheme R^5. Suppose each individual t's true preference
ordering is always in D_t but the authorities cannot be sure of
that. Therefore, when 3 announces R^3 there is no way of determining that she is misrepresenting her preference ordering and thus
majority rule, even on the restricted domain, is manipulable in this
extended sense (see Blin and Satterthwaite, 1976).

A nice example of a voting system that is non-manipulable on a
restricted domain even when players can announce any logically
possible preference ordering is *voting by quota*, proposed and
analysed by Barberà *et al.* (1991).

22. Voting by quota. $T = \{1,2,\ldots,n\}$ and a quota Q is fixed
$(1 \le Q \le n)$. X^* is the set of alternatives. They are not mutually
exclusive so any subset $x \subset X^*$ can be selected by the voting scheme.
The set of possible outcomes is X, the family of subsets of X^*.
Define $g\colon P(X)^T \to X$ by setting $g(p) = \{\alpha \in X^* \mid \alpha \in \max p(t)$ for
at least Q individuals $t \in T\}$.

This voting scheme is manipulable even for $D = L(X)^T$, of
course. For example, set $T = \{1,2,3\}$, $X^* = \{\alpha,\beta,\gamma\}$, $Q = 2$,
and choose $p \in D$ so that $\max p(1) = \{\alpha,\beta\}$, $\max p(2) = \{\alpha\}$,
$\max p(3) = \{\beta\}$ and $(\{\alpha\},\{\alpha,\beta\}) \in Ap(3)$. We have $g(p) =$
$\{\alpha,\beta\}$ by definition: both α and β belong to $\max p(t)$ for two
persons. Person 3 can manipulate g at p via $p(2)$. We have
$g(p') = \{\alpha\}$ if $p'(1) = p(1)$ and $p'(2) = p'(3) = p(2)$. But
$(\{\alpha\},\{\alpha,\beta\}) \in Ap(3)$. However, as Barberà *et al.* demonstrate,
there is a broad family D_1 of preorders with respect to which g is
non-manipulable. (Set $D_t = D_1$ for all t.) The family of *additively
representable* members of $P(X)$ has this property. $R \in P(X)$ is
additively representable if there is a real-valued function u on X^*
such that xRy if and only if $\Sigma_{\alpha \in x} u(\alpha) \ge \Sigma_{\beta \in y} u(\beta)$, with $\Sigma_{\alpha \in x} u(\alpha)$
defined as zero if x is the empty set. The ordering $p(3)$ above
is not additively representable: if $u(\alpha) > u(\alpha) + u(\beta)$ then
$u(\beta) < 0 = u(\varnothing)$ and thus $(\varnothing,\{\beta\}) \in AR$ if R is generated by u.

To show that (D,g) is non-manipulable when D is the set of profiles of additively representable preferences choose any $p \in D$ and
$t \in T$. If $\max p(t) = \varnothing$ then $u(\alpha) < 0$ for all $\alpha \in X^*$. Therefore, it is
not in t's interest to have any alternatives added to $g(p)$ and the best
strategy for t is to declare her true preference scheme $p(t)$. If
$\max p(t) = y \ne \varnothing$ then $u(\alpha) > 0$ for all $\alpha \in y$ if $p(t)$ is generated

by u. Then $y = \{\alpha \in X^* \mid u(\alpha) > 0\}$ and t wants as many members of y as possible added to $g(p)$. Again the best way to ensure this is to declare $p(t)$. There is no R in D_t, or in $P(X)$, that will precipitate an outcome that ranks higher than $g(p)$ in the ordering $p(t)$ given the announcement $p(i)$ by each $i \neq t$. Therefore, (D, g) is non-manipulable in a strong sense.

Although the domain D of additively representable preorders admits a non-manipulable and non-dictatorial voting scheme for any set X^* it does not admit a non-dictatorial social welfare function $f: D \to P(X)$ satisfying the Pareto criterion and Arrow's independence axiom. This follows from Theorem 5.5 and the fact that D has the free triple property, which we now prove.

Let x, y, and z be any three distinct members of X. We have to generate the following preorders on $\{x, y, z\}$ by appropriate choices of utility functions u on X^*: $xPyPz$, $xIyIz$, $xIyPz$, $xPyIz$. (Of course P denotes the asymmetric factor and I the symmetric factor of the desired preorder.) We will define u by specifying $u(\alpha)$ for only one or two members α of X^*. If $u(\lambda)$ is not specified it is understood that $u(\lambda) = 0$. First, suppose that $x \cup y \cup z$ has exactly two members. There are four cases to consider.

Case 1. $xPyPz$. If x is empty and $y = \{\beta\}$ and $z = \{\gamma\}$ then set $u(\beta) = -1$ and $u(\gamma) = -2$. If y is empty and $x = \{\alpha\}$ and $z = \{\gamma\}$ then set $u(\alpha) = 1 = -u(\gamma)$. If z is empty and $x = \{\alpha\}$ and $y = \{\beta\}$ then set $u(\alpha) = 2$ and $u(\beta) = 1$.

Case 2. $xIyIz$. Set $u(\lambda) = 0$ for all λ in X^*.

Case 3. $xIyPz$. If $z = \{\gamma\}$ then set $u(\gamma) = -1$. If z is empty and $x = \{\alpha\}$ and $y = \{\beta\}$ then set $u(\alpha) = 1 = u(\beta)$.

Case 4. $xPyIz$. We know that we can find a utility function u' that generates $zI'yP'x$ (Case 3). Set $u = -u'$.

Now, suppose that we can find the desired utility function whenever $x \cup y \cup z$ has exactly m members, for $2 \leqslant m \leqslant n$. Assume that $x \cup y \cup z$ has $n + 1$ members. Choose any $\alpha \in x \cup y \cup z$ such that $x - \{\alpha\}, y - \{\alpha\}$, and $z - \{\alpha\}$ are distinct subsets of X^*. Then by the induction hypothesis the alternatives $x - \{\alpha\}$, $y - \{\alpha\}$, and $z - \{\alpha\}$ in X can be arbitrarily ordered by appropriate choice of a utility function u on X^*. Then u induces the associated ordering of x, y, and z, with $w \in \{x, y, z\}$ taking the place of $w - \{\alpha\}$.

The discussion of voting by quota has featured a domain on which Arrow's impossibility theorem goes through but the Gibbard-

Satterthwaite theorem does not. There is a domain $D \subset P(X)^T$ with respect to which only dictatorial social welfare functions $f: D \to P(X)$ can satisfy Arrow's independence axiom and the Pareto criterion, but there exist nondictatorial and non-manipulable voting schemes on D (Example 22). On the other hand, Example 5.7 and Theorem 17 show that there are domains on which Arrow's conditions are consistent but the Gibbard–Satterthwaite theorem goes through. That is, there exists a domain $D \subset P(X)^T$ on which one can construct a non-dictatorial social welfare function $f: D \to P(X)$ satisfying the Pareto criterion and Arrow's independence axiom, but on which every non-manipulable voting scheme is dictatorial. (Theorem 17 applies because $L(X) \subset D_t$.) Nevertheless, if one imposes non-dictatorship (of social welfare functions and voting schemes) there is a close relationship between the domains that admit non-manipulable voting schemes and the domains on which one can construct social welfare functions satisfying the Pareto criterion, the independence axiom, and full transitivity of social preference. This has been explored by Maskin (1975*a,b*), Kalai and Muller (1977), and Blin and Satterthwaite (1978).

Part III
Connected Spaces

Introduction

BOTH the Pareto criterion and non-imposition can be dispensed with in proving an impossibility theorem for transitive social preference on a T_1-space that is connected. A richer topological structure has to be assumed if the set of individuals is infinite, but the additional assumptions are elementary and are consistent with virtually every allocation space employed in the literature. In this framework the only non-constant social welfare functions satisfying Arrow's independence axiom are either completely dictatorial or completely inversely dictatorial. The intuition behind the result is simple enough. Suppose that f is not constant. Then there is at least one pair of alternatives over which social preference can vary as the profile of individual preferences varies. By continuity of social preference there must be an *open* set N of alternatives satisfying non-imposition with respect to $f|N$, the restriction of f to that open set. That means that Wilson's theorem (5.4) applies to $f|N$, which cannot be null, and hence must be dictatorial or inversely dictatorial. (Actually, the domain assumption employed in Part III is not strong enough to allow us to use Wilson's theorem directly, but we will choose N so that it has plenty of free triples. Continuity of social preference transmits dictatorship on free triples to dictatorship over all of N.) To grasp the essential idea assume that every triple is free. Suppose that individual t is a dictator for $f|N$. What happens on the boundary of the open set defining $f|N$? Continuity of social preference will force us to extend t's jurisdiction to a neighbourhood of any boundary point of N, and inevitably to the whole space X. Similarly, continuity will extend an inverse dictator's realm to the entire outcome space.

If social preference is merely assumed to be quasitransitive then dictatorship and inverse dictatorship can be avoided. However, the social welfare function will be oligarchical if it satisfies strict non-imposition. Any definition of equity will disquality any social welfare function that is completely independent of some individual's preference scheme, and likewise any rule that inverts someone's preference ordering before taking it into consideration. The only acceptable oligarchy, then, is the one that defines the Pareto aggregation rule: x ranks above y if and only if *everyone* strictly prefers x to y. Even without the Pareto criterion, we are led back to Pareto optimality as the only welfare standard consistent with the axioms. Whether social preference is fully transitive or merely quasitransitive, a relaxation of the Pareto criterion does not open the door to an acceptable rule for settling distributional disputes. If the social welfare function is authoritarian then distributional issues are 'resolved' but only in the most unsatisfactory way imaginable. If the Pareto aggregation rule is used there is complete silence on distributional matters.

8 Impossibility without Efficiency

THIS chapter considers both transitive and quasitransitive social preference. When social preference is fully transitive we obtain a strong impossibility theorem without any vestige of efficiency. This is because continuity of social preference disqualifies all but constant or authoritarian social welfare functions. Of course Arrow's independence axiom is in force. The case of quasitransitive social preference is examined within the ambit of the strict non-imposition condition as well. Undoubtedly, stronger results await discovery. Weak versions of the free-triple property are assumed in this chapter, but first we investigate the implications of full transitivity and the free-triple property then use the results to prove a theorem for the unrestricted domain case. The reader who is only interested in results that will later apply to economic environments (Chapters 10–13) can skip to Definition 15.

Recall that $x \geq_f y$ means that (x,y) belongs to $f(p)$ for some $p \in D$ (4.1). For arbitrary $x \in X$ let $G(x) = \{y \in X \mid x \geq_f y \geq_f x\}$. We will prove that $f \mid G(x)$ satisfies non-imposition (3.3). $G(x)$ is called a *component* of f. (Recall that $f \mid Y$ is the restriction of f to Y. See Definition 16 below.)

1. Lemma. Let T be any set. If $f: D \to P(X)$ satisfies Arrow's independence axiom and D has the free-triple property then $\geq_f$ is transitive. Consequently, $f \mid G(x)$ satisfies non-imposition for every $x \in X$.

Proof. Suppose that $x \geq_f y \geq_f z$. We certainly have $x \geq_f z$ if the three alternatives are not distinct. Suppose, then, that $x \neq y \neq z \neq x$. By definition, there are profiles $p, p' \in D$ such that $(x,y) \in f(p)$ and $(y,z) \in f(p')$. By the free-triple property there is a profile $p'' \in D$ such that $p''(t) \cap \{x,y\}^2 = p(t) \cap \{x,y\}^2$ for all $t \in T$, and $p''(t) \cap \{y,z\}^2 = p'(t) \cap \{y,z\}^2$ for all $t \in T$. Then $(x,y), (y,z) \in f(p'')$ by the independence axiom and thus $(x,z) \in$

$f(p'')$ by transitivity of $f(p'')$. Therefore, $x \geq_f z$ by definition. If y and z belong to $G(x)$ then $y \geq_f x \geq_f z$ and $z \geq_f x \geq_f y$ and therefore $y \geq_f z \geq_f y$ by transitivity of $\geq_f$. ∎

2. Lemma. Let T be any set. If $f: D \to P(X)$ satisfies Arrow's independence axiom and D has the free-triple property then for all $x,y \in X$, $G(x) \cap G(y) \neq \varnothing$ implies $G(x) = G(y)$.

Proof. Suppose that z belongs to $G(x)$ and $G(y)$. Then $x \geq_f z \geq_f y$ and $y \geq_f z \geq_f x$. Therefore, $x \geq_f y$ and $y \geq_f x$ by Lemma 1. For any $v \in G(x)$ we have $v \geq_f x \geq_f y$ and $y \geq_f x \geq_f v$. Therefore, $v \geq_f y$ and $y \geq_f v$ by Lemma 1 and hence $G(x) \subset G(y)$. Similarly, $G(y) \subset G(x)$. ∎

For arbitrary subsets Y and Z of X write $Y \gg_f Z$ if $(y,z) \in Af(p)$ holds for all $y \in Y$, $z \in Z$, and $p \in D$. Lemma 3 proves that $G(x) \gg_f G(y)$ or $G(y) \gg_f G(x)$ must hold unless $G(x) = G(y)$.

3. Lemma. Let T be any set. If $f: D \to P(X)$ satisfies Arrow's independence axiom and D has the free-triple property then for all $x,y \in X$ either $G(x) = G(y)$ or $G(x) \gg_f G(y)$ or $G(y) \gg_f G(x)$.

Proof. Suppose that $v,v' \in G(x)$, $w,w' \in G(y)$, $v \geq_f w$ and $w' \geq_f v'$. Then $x \geq_f v \geq_f w \geq_f y$ so $x \geq_f y$, applying Lemma 1 twice. Similarly, $y \geq_f w' \geq_f v' \geq_f x$ and hence $y \geq_f x$. Therefore, $x \in G(y)$ and hence $G(x) \cap G(y) \neq \varnothing$. Thus, $G(x) = G(y)$ by Lemma 2. ∎

A social welfare function is *authoritarian* if it is dictatorial or inversely dictatorial. We next prove that $G(x)$ is closed if $f \mid G(x)$ is authoritarian.

4. Lemma. Let T be any set and let G be any component of f. If $f: D \to P(X)$ satisfies Arrow's independence axiom, D has the free-triple-property, and $f \mid G$ is authoritarian then G is closed.

Proof. Suppose that t is a dictator for $f \mid G$ and G is not closed. Then there is some $y \notin G$ in the closure of G. Choose any $x \in G$ and set $G' = G(y)$. Assume that $G \gg_f G'$. By the free-triple property there is some $p \in D$ such that $(y,x) \in Ap(t)$. Because $(x,y) \in Af(p)$ there are neighbourhoods $N(x)$ and $N(y)$ of x and y respectively such that $(v,w) \in Af(p)$ for all $v \in N(x)$ and $w \in N(y)$. Because $(y,x) \in Ap(t)$ there are neighbourhoods $N'(y)$ and $N'(x)$ of y and x respectively such that $(w',v') \in Ap(t)$ for all $w' \in N'(y)$ and $v' \in N'(x)$. Because y is in the closure of G there is some $w \in N(y)$

$\cap N'(y) \cap G$. We have $(w,x) \in Af(p)$ because $w \in N'(y)$ and t is a dictator for $f|G$. And $(x,w) \in Af(p)$ because $w \in N(y)$, an obvious contradiction. Therefore, $G \gg_f G'$ can be ruled out. But $G \neq G'$ and thus $G' \gg_f G$ by Lemma 3. But this will lead to a contradiction by an analogous argument. Therefore G must be closed. Similarly, we can show that G is closed if t is an inverse dictator for $f|G$. ■

Now we show that f itself is authoritarian if $f|G$ is authoritarian for some component G.

5. Lemma. Let T be any set, and assume that X is a connected topological space. If $f: D \to P(X)$ satisfies Arrow's independence axiom and D has the free-triple property then $G = X$ if G is a component of f and $f|G$ is authoritarian and G has at least two members.

Proof. Suppose that $G \neq X$, G contains two or more elements, and t is a dictator for $f|G$. G is closed by Lemma 4 and X is connected so $X - G$ is not closed. (If Y and Z are disjoint and non-empty closed subsets of X then $X - Y$ and $X - Z$ are open. If $Y \cup Z = X$ then $X - Y = Z$, $X - Z = Y$ and X is the union of two disjoint and non-empty open sets, contradicting connectedness of X.) Because $X - G$ is not closed the set G contains some y in the closure of $X - G$. $G(y) = G$ by Lemma 2. Choose some $z \in G(y)$ such that $z \neq y$. Now choose $p,p' \in D$ such that $(y,z) \in Ap(t)$ and $(z,y) \in Ap'(t)$. This is possible by the free-triple property. We have $(y,z) \in Af(p)$ and $(z,y) \in Af(p')$ because t dictates $f|G$. Then there are neighbourhoods $N(y)$ and $N'(y)$ of y such that $(v,z) \in Af(p)$ for all $v \in N(y)$ and $(z,v') \in Af(p')$ for all $v' \in N'(y)$. But y is in the closure of the set $X - G$ so there is some $v \in N(y) \cap N'(y) \cap (X - G)$. Obviously, $v \notin G$ so $G \neq G(v)$. Then $G(v) \gg_f G$ does not hold because $z \in G$ and $(z,v) \in Af(p')$. And $G \gg_f G(v)$ is false because $(v,z) \in Af(p)$. We have contradicted Lemma 3. Therefore, G must equal X.

If t is an inverse dictator for $f|G$ proceed as above. We will have $(z,y) \in Af(p)$ and $(y,z) \in Af(p')$ because t is an *inverse* dictator. Again, we can find some v such that $G(v) \neq G$ with $(z,v) \in Af(p)$ and $(v,z) \in Af(p')$, contradicting Lemma 3. ■

6. Lemma. Let T be any set. Assume that X is a connected topological space and that $f: D \to P(X)$ satisfies Arrow's independence axiom and D has the free-triple property. If f is not constant then there is some component G such that G contains an open set and $f|G$ is not constant.

Proof. If f is not constant there exist $x,y \in X$ and $p,p' \in D$ such that $(x,y) \in Af(p)$ and $(y,x) \in f(p')$. Recall Proposition 2.19. Because X is connected and $f(p)$ is complete, transitive, and continuous there is some $z \in X$ such that $(x,z) \in Af(p)$ and $(z,y) \in Af(p)$. By continuity of $f(p)$ there is some neighbourhood $N(z)$ of z such that $(x,v),(v,y) \in Af(p)$ for all $v \in N(z)$. We also have $y \geq_f x$. Then, $v \geq_f y \geq_f x$ for all $v \in N(z)$ and thus $v \geq_f x$ by Lemma 1. Therefore, $N(z) \subset G(x)$. The social welfare function $f|G(x)$ is not constant because $(x,v) \in Af(p)$ and $v \geq_f x$. ∎

We have not yet assumed that X has at least three members. This will be implicit in our first theorem which assumes that the topology on X is T_1. (If $x \neq y$ there is some open set containing x but not y.) Obviously, a non-singleton connected T_1-space cannot even be finite (Proposition 2.5). We also need to assume a finite society in order to invoke Wilson's theorem (5.4). A simple topological lemma precedes the theorem.

7. Lemma. A singleton subset of a T_1-space is closed. And if X is a connected T_1-space with more than one member then every non-empty open subset of X is infinite.

Proof. Suppose $y \neq x$. Then there is an open set containing y but not x. Then y cannot be in the closure of $\{x\}$. Therefore $\{x\}$ is closed.

If X is a T_1-space then $X - \{x\}$ is open for any $x \in X$ because $\{x\}$ is closed. If $Y \subset X$ then $X - Y = \cap \{X - \{x\} | x \in Y\}$. Therefore, if Y is finite the set $X - Y$ is open as the finite intersection of open sets. If Y is also open then X cannot be connected. (If $X = Y$ and Y is finite then X is not connected by Proposition 2.5.) ∎

8. Theorem. Let T be any finite set. Assume that X is a connected T_1-space. If D has the free-triple property and $f: D \to P(X)$ satisfies Arrow's independence axiom then f is either constant or authoritarian.

Proof. If X is a singleton then f is certainly constant. Assume that X has more than one member. Then any open subset of X is infinite by Lemma 7.

By Lemma 6 there is some $x \in X$ such that $f \mid G(x)$ is not constant and $G(x)$ contains an open (and hence infinite) set. Therefore, $f \mid G(x)$ is a non-null social welfare function satisfying Arrow's independence axiom and non-imposition. Obviously, $D(G(x))$ has the free-triple property. Therefore, $f \mid G(x)$ is either dictatorial or inversely dictatorial by Theorem 5.4. As a result, $G(x) = X$ by Lemma 5 and hence $f = f \mid X = f \mid G(x)$ is authoritarian. ■

As one would expect, continuity of social preference implies that a dictatorship must be a complete dictatorship and an inverse dictatorship must also be complete. To prove this we need an additional assumption, however. It is one that is mild enough to be satisfied in every economic application where continuity of individual preference is a natural property. Our new assumption is referred to as *regularity*. The reader is reminded that the domain is always assumed to be a product set, $D = \Pi_{t \in T} D_t$.

9. Regularity. Let X be a topological space and let D be a domain for X. The domain $D \subset P(X)^T$ is regular if for all $t \in T$, $R \in D_t$, and $x,y \in X$ such that $(x,y) \in SR$ there exists some $R' \in D_t$ such that $(x,y) \in SR'$ and every neighbourhood of x contains a point z such that $(z,x) \in AR'$.

One can quickly verify that W, the family of profiles of economic preferences on Ω, is regular. Set $R' = R$ and exploit the monotonicity of R. Recall that $B(X)$ is the family of complete and continuous binary relations on X. The following completeness proposition does not require any transitivity of social preference, nor does it depend on the specification of a topology, although the choice of a topology has a bearing on whether D qualifies as regular.

10. Proposition. Let T be any set and let X be any topological space. If D is regular and $f: D \to B(X)$ satisfies Arrow's independence axiom then f is completely authoritarian if it is authoritarian, and it is completely oligarchical if it is oligarchical.

Proof. If t is a dictator for f then $(\{t\}, \varnothing)$ is an oligarchy. Similarly, $(\varnothing, \{t\})$ is an oligarchy if t is an inverse dictator. Therefore, it is only necessary to prove the oligarchy part of the theorem.

Suppose that (I,J) is an oligarchy for f and $(x,y) \in Af(p)$ but $(x,y) \in Sp(t)$ for some $t \in I$. Choose $p' \in D$ such that $p'(i) = p(i)$

for all $i \neq t$, $(x,y) \in Sp'(t)$ and every neighbourhood of y contains at least one point z such that $(z,y) \in Ap'(t)$. The profile p' exists by regularity. We have $(x,y) \in Af(p')$ by the independence axiom. Because $f(p')$ is continuous there is a neighbourhood $N(y)$ of y such that $(x,z) \in Af(p')$ for all $z \in N(y)$. But we have $(z,y) \in Ap'(t)$ for some $z \in N(y)$ by choice of p', and thus $(z,x) \in Ap'(t)$, contradicting the fact that t has veto power. Similarly, if $t \in J$ and $(x,y) \in Af(p) \cap Sp(t)$ we have $N(x) \times \{y\} \subset Af(p')$ for some neighbourhood $N(x)$ of x and some $p' \in D$, although $(z,x) \in Ap'(t)$ for some $z \in N(x)$ and hence $(z,y) \in Ap'(t)$, contradicting the fact that t has an inverse veto. Therefore, an oligarchy must be complete. ∎

Before developing a result that applies to economic environments we prove that the unrestricted domain $P(X)^{\mathrm{T}}$ has the free-triple property and satisfies the hypothesis of Proposition 10. A richer topological structure must be employed. X is assumed to be a normal and connected topological space satisfying the first axiom of countability. (Any connected metric space has this property. See Bourbaki 1966.) Three topological lemmas are required. The first proves that a finite subset of X is a G_δ set. (Recall that a G_δ set is one that can be expressed as the intersection of a countable collection of open sets.)

11. Lemma. If X is a T_1-space satisfying the first axiom of countability then any finite subset of X can be expressed as the intersection of a countable collection of open sets.

Proof. For $x \in X$ let $\{x\} = \cap \{N_i(x) \mid i \geq 1\}$. (Proposition 2.7.) Set $M_i(x) = N_1(x) \cap N_2(x) \cap \ldots \cap N_i(x)$. Therefore, $\{x\} = \cap \{M_i(x) \mid i \geq 1\}$. Given the subset Y of X let $N_i^* = \cup \{M_i(x) \mid x \in Y\}$. Each set N_i^* is open as the union of open sets. If $z \notin Y$ then for $x \in Y$ we have $z \neq x$ and hence $z \notin M_{i(x)}(x)$ for some finite integer $i(x)$. Because the sets $M_i(x)$ are decreasing we have $z \notin M_j(x)$ for all $j \geq i(x)$. If Y is finite we can choose $h = \max\{i(x) \mid x \in Y\}$. Then $z \notin M_j(x)$ for all $j \geq h$ and all $x \in Y$. Therefore, $z \notin \cap \{N_i^* \mid i \geq 1\}$ and hence $Y = \cap \{N_i^* \mid i \geq 1\}$. ∎

The next lemma is a consequence of Urysohn's theorem on the existence of non-constant continuous functions (Corollary 2.11). It proves that any preorder on a finite subset of the space is contained in some *continuous* preorder on the space. A modest topological structure is required.

12. Lemma. Let X be a normal T_1-space satisfying the first axiom of countability. If Y is a finite subset of X and R is any preorder on Y there is some $R' \in P(X)$ such that $R' \cap Y^2 = R$.

Proof. A preorder R on a finite set Y can be represented by means of a partition $Y_1, Y_2, \ldots Y_m$ of Y. We have xRy if and only if $x \in Y_i$, $y \in Y_j$, and $i \geq j$. By Lemma 11 and Corollary 2.11 for each i there is a continuous function $g_i: X \to [0,1]$ such that $\cup \{ Y_j | i \leq j \leq m \} = \{ x \in X | g_i(x) = 1 \}$ and $\cup \{ Y_j | 1 \leq j \leq i - 1 \} = \{ x \in X | g_i(x) = 0 \}$. Set $g = g_1 + g_2 + \ldots + g_m$. Then $g(x) = i$ if $x \in Y_i$. And g is a continuous function as the finite sum of continuous functions. Set $xR'y$ if and only if $g(x) \geq g(y)$. Then R' belongs to $P(X)$ and $R' \cap Y^2 = R$. ∎

13. Lemma. Let T be any set. If X is a normal and connected T_1-space satisfying the first axiom of countability then $P(X)^T$ is regular.

Proof. Let x and y be distinct members of X. Both $\{x\}$ and $\{y\}$ are closed by Lemma 7. Therefore, their union $\{x,y\}$ is closed, and by Corollary 2.11 there is a continuous function $g: X \to [0,1]$ such that $g(x) = g(y) = 0$ and $g(z) \neq 0$ if $x \neq z \neq y$. Define R by setting $(v,w) \in R$ if and only if $g(v) \geq g(w)$. Then R is a continuous preorder on X. Let $N(x)$ be any neighbourhood of x. An open set in a connected T_1-space cannot be finite (Lemma 7). Therefore, there is some $z \in N(x)$ such that $x \neq z \neq y$. Then $(z,x) \in AR$ and $(x,y) \in SR$. ∎

Obviously, Lemma 12 proves that $P(X)^T$ has the free-triple property. Therefore, we can provide a *characterization* of Arrow's independence axiom.

14. Theorem. Let T be any finite set and let X be a normal and connected T_1-space satisfying the first axiom of countability. Then $f: P(X)^T \to P(X)$ satisfies Arrow's independence axiom if and only if it is constant or completely authoritarian.

Proof. (i) A constant or completely dictatorial social welfare function satisfies Arrow's independence axiom for any specification of the domain and topology. The same can be said of a completely inversely dictatorial social welfare function.

(ii) Suppose that f satisfies Arrow's independence axiom and the rest of our hypothesis. Then the domain has the free-triple property by Lemma 12. Therefore, f is either constant or authoritarian by

Theorem 8. If f is authoritarian it is completely authoritarian by Proposition 10 and Lemma 13. ■

Most of the topological apparatus behind Theorem 14 is used to prove that any complete and transitive relation on an arbitrary finite subset of X is contained in some continuous (individual) preorder on X. If there is prior information to guarantee that arbitrary triples from X can be arbitrarily ordered then we merely need to assume that X is a connected T_1-space (Theorem 8). This would qualify the Riesz spaces used by Aliprantis and Brown (1983) to study the existence of equilibrium with an infinite number of commodities, and the spaces of Borel probability measures on a compact real interval used in the study of decision-making under uncertainty (e.g. Allen 1987; Machina 1982). Any connected metric space is normal and first-countable (Bourbaki 1966, ii.182). Therefore, most of the standard allocation spaces satisfy the stronger hypothesis of Theorem 14 and so do the spaces of sigma-algebras of measurable subsets of a non-atomic measure space used by Berliant (1984) to study economies in which land is realistically modelled, and the Mackey topology employed by Bewley (1972) in the study of equilibrium with an infinite number of commodities, and by Brown and Lewis (1981) to study infinite horizon choice.

A number of papers examine the implications of continuity of the social welfare function itself. Chichilnisky (1982*a*) represents individual and social preference as vector fields and proves the *topological* equivalence of complete dictatorship and the requirement that the social welfare function be continuous and satisfy the Pareto criterion. The result is very striking because the independence axiom is not imposed, but it is difficult to know what ethical significance to attach to the property of topological equivalence to dictatorship. Theorem 14 establishes the *logical* equivalence of complete dictatorship and the independence axiom within the family of non-constant social welfare functions.

Again using the vector field approach Chichilnisky (1980, 1982*b*) proves that there is no continuous social welfare function satisfying anonymity (symmetry across individuals) and unanimity (if all individuals have the ordering R then the social preference is R). Le Breton and Uriarte (1990) demonstrate that this result depends crucially on the choice of topology for the space of preferences, although Baigent and Huang (1990) should be consulted for a definitive discussion of this issue. On the other hand, Theorem 14 applies to any topological space of alternatives that economists are likely to employ.

Continuous social welfare functions are also examined in McManus (1982) and Ferejohn and Packel (1983). They prove that the independence axiom, the Pareto criterion, and continuity of the social welfare function are inconsistent. There is a sense in which continuity of the social ordering is implicit (Ferejohn and Packel: 67) so our theorem is substantially more general.

Ferejohn *et al.* (1980) assume only continuity of social preference and show how this assumption implies an underlying structure of winning and losing coalitions. Not everyone will find their other assumptions to be intuitively basic, but their result is remarkable because the social preference relation is not required to posses any degree of transitivity.

A variant of Arrow's impossibility theorem emerges as a corollary of Theorem 14. If $f\colon P(X)^{\mathrm{T}} \to P(X)$ is a non-constant social welfare function satisfying the independence axiom and there is one pair of alternatives x and y such that x ranks at least as high as y in the social preference scheme for at least one situation in which everyone strictly prefers x to y then there is one individual whose preference ordering always equals the social ordering. (Continuity of social preference must be imposed, of course.) Murakami (1961) pointed out that *local* dictatorship is implied by the original Arrow (1951) conditions. We have established complete dictatorship, a consequence of the topological desiderata.

Now we develop a result that will be instrumental in proving Theorem 8 for the domain W of classical economic preferences on the space Ω of allocations of private goods. We first preview the argument that will be developed in Chapter 10. Assume that there are two private goods ($k = 2$). The key is to apply Theorem 8 to connected subspaces of Ω and then use intersecting subspaces to extend the conclusion to the overall social welfare function $f\colon W \to P(\Omega)$. For example, set $Y_t = \{x \in \mathbf{E}^2_{++} \mid \log x_1 + \log x_2 = 0\}$ and $Y = \Pi_{t \in T} Y_t$. Then Y is connected as the product of connected spaces (Propositions 2.14 and 2.18). Individual t's preference scheme depends only on $x(t)$. Because $Y_t = \{(a,b) \in \mathbf{E}^2_{++} \mid ab = 1\}$, a rectangular hyperbole, every triple $\{x,y,z\}$ from Y_t can be arbitrarily ordered by choosing an appropriate classical preference ordering on $\mathbf{E}^2_+$. That is, for every $R \in P(\{x,y,z\})$ there is some classical utility function that includes R. In spite of this Y

does *not* have the free-triple property. Consider $x,y,z \in Y$ with $x(t) = (2,\frac{1}{2})$ for all t, $y(1) = z(1) = (2,\frac{1}{2})$, and for all $t > 1$, $y(t) = (1,1)$ and $z(t) = (\frac{1}{2},2)$. Then $x \neq y \neq z \neq x$ and $\{x,y,z\}$ is a three-element set. But $x(1) = y(1) = z(1)$ so $W_1(\{x,y,z\})$ is a singleton $\{R\}$ with (x,y), $(y,z) \in SR$. (Recall that $D_t(Y)$ is the set of preorders $R \cap Y^2$ on Y such that $R \in D_t$.) Obviously, $W_1(\{x,y,z\}) \neq P(\{x,y,z\})$. However, $W(Y)$ *almost* has the free-triple property. For any three elements $x,y,z \in Y$ there will be three alternatives x',y',z' with each point arbitrarily close to its unprimed relative and such that $W(\{x',y',z'\})$ has the free-triple property. This relationship is pursued in more detail in Chapter 10. Our aim here is to introduce and motivate a weaker domain assumption. (Verify that $W(Y)$ satisfies Definition 15 below.) First, we introduce some notation.

Given the domain D we let $F_t(x)$ denote the set of outcomes $y \neq x$ such that $D_t(\{x,y\}) = P(\{x,y\})$. Let $F(x) = \bigcap_{t \in T} F_t(x)$. We say that the pair $\{x,y\}$ is *free* if $y \in F(x)$, and we say that arbitrary $Y \subset X$ is free if $D(Y) = P(Y)^T$ and Y is not a singleton.

15. Standard domain. D is a standard domain if it has the following four properties.

15a If $x \notin F_t(y)$ then $D_t(\{x,y\}) = \{R^0\}$, where $R^0 = \{x,y\}^2$, the null relation on $\{x,y\}$.

15b If $x \in F_t(y)$, $y \in F_t(z)$, and $z \in F_t(x)$ then $D_t(\{x,y,z\}) = P(\{x,y,z\})$.

15c For any $x,y \in X$ the set $F(x) \cap F(y)$ is not empty.

15d If N and M are non-empty open sets and $\{x,y\}$ and $\{a,b\}$ are free then there exists $w \in N$ and $z \in M$ such that $\{x,y,w\}$, $\{y,w,z\}$, $\{w,z,a\}$, and $\{z,a,b\}$ are free triples.

Conditions 15c and 15d imply that D has the following property:

15e If $x \in X$ and N is a non-empty open set then there exists $y \in N$ such that $\{x,y\}$ is free.

To prove this, choose any $x \in X$. There is some $x' \in X$ such that $\{x,x'\}$ is free by 15c. Then $\{x,x',y\}$ is free for some $y \in N$ by 15d. Note that 15c and 15d also imply the existence of a free triple.

Condition 15a says that person t is always indifferent between x and y unless all possible orderings of x and y are embodied in D_t. Condition 15b says that if for each pair of distinct alternatives from $\{x,y,z\}$ each ordering of that pair is included in some member of

D_t then each ordering of $\{x,y,z\}$ itself is included in some member of D_t. This is not implied by the definition of F_t. For example, $X = \{x,y,z\}$ and D_t consist of the following three preorders represented as three columns:

$$
\begin{array}{ccc}
x & y & z \\
yz & zx & xy
\end{array}
$$

(In each case two alternatives are indifferent to each other and they rank strictly below the third alternative.) We have $x \in F_t(y)$, $y \in F_t(z)$, and $z \in F_t(x)$ but D_t certainly does not contain all preorders on $\{x,y,z\}$. (Ten are missing.) Although there is no guarantee that every triple is free, condition 15c requires that for each pair $\{x,y\}$ there exists an alternative z such that the pairs $\{x,z\}$ and $\{y,z\}$ are both free. Note that $\{x,y,z\}$ may not be free. Note that 15c implies that there exist at least two free pairs if X contains at least two alternatives. Condition 15d says that any two free pairs $\{x,y\}$ and $\{a,b\}$ can be linked by four overlapping sets of free triples. Condition 15e says that a free pair can always be obtained by an appropriate choice of points from two open sets.

Any domain with the free-triple property is standard if X is a (non-singleton) connected T_1-space. The definition of a standard domain could be modified so that it is logically weaker that the free-triple property without invalidating any of the proofs to follow. For example, 15d could be waived whenever $N - \{x,y\}$ or $M - \{a,b\}$ contained fewer than two alternatives. We have chosen to sacrifice a little generality to streamline the exposition.

For $D \subset P(X)^T$ and $Y \subset X$ the subdomain $D(Y)$ is the set of profiles $p \in P(Y)^T$ such that there is some $q \in D$ satisfying $p(t) = q(t) \cap Y^2$ for all $t \in T$. If $f: D \to B(X)$ satisfies Arrow's independence axiom and Y is a subset of X we can define $f|Y$, the restriction of f to Y.

16. Definition. If $f: D \to B(X)$ satisfies Arrow's independence axiom and $Y \subset X$ then $f|Y$ is the social welfare function mapping $D(Y)$ into $B(Y)$ and satisfying $f|Y(p) = f(q) \cap Y^2$ for any $q \in D$ such that $q(t) \cap Y^2 = p(t)$ for all $t \in T$.

Now we prove nine lemmas leading up to the key theorem (26) of this chapter.

17. Lemma. Let T be any set. If $f: D \to P(X)$ satisfies Arrow's independence axiom and D is standard then $x \geq_f y \geq_f z$ and

$x \in F(z)$ imply $x \geq_f z$, with $x >_f z$ actually holding if $x >_f y$ or $y >_f z$.

Proof. Suppose $(x,y) \in f(p)$ and $(y,z) \in f(p')$. Define $p'' \in D$: If $x \notin F_t(y)$ set $p''(t) = p'(t)$. Then $p''(t) \cap \{x,y\}^2 = p(t) \cap \{x,y\}^2$ because $D_t(\{x,y\})$ is a singleton by 15a. If $y \notin F_t(z)$ set $p''(t) = p(t)$. In either case we have $p''(t) \cap \{x,y\}^2 = p(t) \cap \{x,y\}^2$ and $p''(t) \cap \{y,z\}^2 = p'(t) \cap \{y,z\}^2$. If $x \in F_t(y)$ and $y \in F_t(z)$ then $D_t(\{x,y,z\}) = P(\{x,y,z\})$ by property 15b because $x \in F(z)$ by hypothesis. Then there exists $p''(t) \in D_t$ such that $p''(t) \cap \{x,y\}^2 = p(t) \cap \{x,y\}^2$ and $p''(t) \cap \{y,z\}^2 = p'(t) \cap \{y,z\}^2$. Therefore, $(x,y) \in f(p'')$ and $(y,z) \in f(p'')$ by Arrow's independence axiom. Thus, $(x,z) \in f(p'')$ by transitivity and thus $x \geq_f z$ by definition. If (x,y) belongs to $Af(p)$ or (y,z) belongs to $Af(p')$ then either (x,y) or (y,z) belongs to $Af(p'')$ in which case we actually have $(x,z) \in Af(p'')$ and hence $x >_f z$. ■

18. Lemma. Let T be any set. Assume that X is a connected T_1-space. If $f: D \to P(X)$ is a non-constant social welfare function satisfying Arrow's independence axiom and D is standard then there exist $x,y \in X$ such that $x >_f y >_f x$ and $\{x,y\}$ is free.

Proof. If f is not constant there exist $x,z \in X$ and $p,p' \in D$ such that $(x,z) \in Af(p) \cap -f(p')$. Because X is connected and $f(p)$ is continuous there is some $w \in X$ such that $(x,w),(w,z) \in Af(p)$. (Lemma 2.19) And there is some neighbourhood $N(w)$ of w such that $\{x\} \times N(w)$ and $N(w) \times \{z\}$ are contained in $Af(p)$. By 15e, $N(w) \cap F(x) \neq \emptyset$. Choose $y \in N(w) \cap F(x)$. We have $y >_f z \geq_f x$ and thus $y >_f x$ by Lemma 17. Then $x >_f y >_f x$. ■

Consider the free pair $\{x,y\}$ of Lemma 18. There exist profiles p and p' such that $(x,y) \in Af(p)$ and $(y,x) \in Af(p')$. By continuity there exist neighbourhoods $N(x)$, $N(y)$, $N'(x)$, $N'(y)$ such that $N(x) \times N(y) \subset Af(p)$ and $N'(y) \times N'(x) \subset Af(p')$. Set $N = N(x) \cap N'(x)$ and $M = N(y) \cap N'(y)$. By 15e we can choose $x^* \in N$ and then $y^* \in M$ such that $\{x^*,y^*\}$ is free. By 15d there is some $z^* \in M$ such that $\{x^*,y^*,z^*\}$ is a free triple. These sets and alternatives remain fixed until we reach Theorem 26. (If X is a connected T_1-space then N and M are infinite if X is not a singleton.) Let $Z = \{x^*,y^*,z^*\}$. It is implicit in Lemmas 19–25 that X is a connected T_1-space, f is non-constant and satisfies Arrow's independence axiom, and D is standard.

19. Lemma. Let T be any set. If $x \in N$, $y \in M$, $z \in N \cup M$, and $\{x,y,z\}$ is a free triple then $a >_f b$ holds for all $a,b \in \{x,y,z\}$ such that $a \neq b$.

Proof. If $\{a,b\} \cap N \neq \{a,b\} \neq M \cap \{a,b\}$ then the conclusion follows from the choice of N and M. If, say, $a,b \in N$ then $c \in M$ for $c \in \{x,y,z\} - \{a,b\}$. Then $a >_f c >_f b$ and hence $a >_f b$ by Lemma 17. A similar argument leads to the same conclusion if a and b both belong to M. ∎

At this point it is necessary to assume that T is finite.

20. Lemma. $f|Z$ is dictatorial or inversely dictatorial if T is finite.

Proof. $f|Z$ is non-constant and satisfies non-imposition by Lemma 19. Then $f|Z$ is authoritarian by Theorem 5.4. ∎

For concreteness assume that person 1 is a dictator for $f|Z$. We maintain this assumption for the next five lemmas. The case where $f|Z$ is inversely dictatorial is handled by means of the obvious counterparts to the following lemmas.

21. Lemma. 1 dictates $f|\{x^*,y\}$ for all $y \in N \cap F(x^*)$ if T is finite.

Proof. 1 is a dictator for $f|\{x^*,y^*\}$. Choose $w \in N$ and $z \in M$ such that $\{x^*,y^*,w^*\}$, $\{y^*,w,z\}$, $\{w,z,x^*\}$, and $\{z,x^*,y\}$ are free triples (15d). For each such triple Y the social welfare function $f|Y$ is non-constant and satisfies non-imposition by Lemma 19. Then $f|Y$ is authoritarian by Theorem 5.4. Then person 1 is a dictator for $f|\{x^*,y^*,w\}$ because 1 dictates $f|\{x^*,y^*\}$. Then 1 dictates each $f|Y$ because each pair of adjacent triples has two members in common. This proves that 1 dictates $f|\{x^*,y\}$. ∎

22. Lemma. For arbitrary $x \in N$ and all $y \in N \cap F(x)$ individual 1 is a dictator for $f \mid \{x,y\}$ if T is finite.

Proof. Choose $z \in M$ such that $\{x,y,z\}$ is free (15e and 15d). The social welfare function $f \mid \{x,y,z\}$ is non-null and satisfies non-imposition by Lemma 19. Then there is a dictator or an inverse dictator for $f \mid \{x,y,z\}$ by Theorem 5.4.

Suppose that $t \neq 1$ is a dictator for $f \mid \{x,y,z\}$. Choose $p \in D$ such that $(x,z) \in Ap(t) \cap - Ap(1)$. Then $(x,z) \in Af(p)$. There exist neighbourhoods $N(x)$ and $N(z)$ such that $N(x) \times N(z) \subset [Af(p) \cap - Ap(1)]$. Choose $x' \in N(x) \cap N$ and $z' \in N(z) \cap M$ such that $\{x^*,x',z'\}$ is free. Then 1 dictates $f \mid \{x^*,x',z'\}$ because $f \mid \{x^*,x',z'\}$ is either dictatorial or inversely dictatorial by Lemma

19 and Theorem 5.4, and 1 must be the dictator by Lemma 21. This implies $(z',x') \in Af(p)$, a contradiction. The case where some t is an inverse dictator for $f \mid \{x,y,z\}$ can be ruled out by the analogous argument. Therefore, 1 must be a dictator for $f \mid \{x,y,z\}$. ∎

Now, let Y denote the set of $x \in X$ such that 1 is a dictator for $f \mid \{x,y\}$ for some $y \in N \cap F(x)$.

23. Lemma. If $x \in Y$ and $y \in Y \cap F(x)$ then 1 is a dictator for $f \mid \{x,y\}$ if T is finite.

Proof. If x and y belong to Y and $x \neq y$ then there exist x' and y' in N such that 1 dictates $f \mid \{x,x'\}$ and $f \mid \{y,y'\}$ and both $\{x,x'\}$ and $\{y,y'\}$ are free. Then we have $x >_f x' >_f x$ and $y >_f y' >_f y$. Therefore there exist open subsets $N(x')$ and $N(y')$ of N containing x' and y' respectively and such that $x >_f v >_f x$ and $y >_f w >_f y$ for all $v \in N(x')$, $w \in N(y')$. Choose v and w so that $\{x,y,v\}$ and $\{y,v,w\}$ are free triples (15d). Then 1 dictates $f \mid \{v,w\}$ by Lemma 22. Therefore, $v >_f w >_f v$. Now we have $x >_f v >_f w >_f y >_f w >_f v >_f x$ and hence both $f \mid \{y,v,w\}$ and $f \mid \{x,y,v\}$ satisfy non-imposition by Lemma 17. Therefore, 1 dictates $f \mid \{y,v,w\}$ by Theorem 5.4 and the fact that 1 dictates $f \mid \{v,w\}$. Thus 1 dictates $f \mid \{x,y,v\}$ by Theorem 5.4 and the fact that 1 dictates $f \mid \{y,v\}$. ∎

24. Lemma. Y is closed if T is finite.

Proof. Suppose that y belongs to the closure of Y. Choose $x \in N \cap F(y)$. (The set is not empty by 15e.) Suppose $(y,x) \in Af(p)$ for all $p \in D$. Choose $p \in D$ such that $(x,y) \in Ap(1)$. We have $(y,x) \in Af(p)$. There exist neighbourhoods $N(x)$ and $N(y)$ such that $N(y) \times N(x) \subset [Af(p) \cap - Ap(1)]$. Choose $y' \in N(y) \cap Y$. By 15c there is some $y'' \in X$ such that $\{y',y''\}$ is free and hence by 15d we can find some $x' \in N(x) \cap N \cap F(y')$. We have $x' \in Y$ by Lemma 22. Then 1 dictates $f \mid \{x',y'\}$ by Lemma 23. Then $(x',y') \in Af(p)$, a contradiction. Similarly, we can rule out the possibility that $x \in N \cap F(y)$ and $(x,y) \in Af(p)$ holds for all $p \in D$. Now choose $x,z \in N$ such that $\{x,y,z\}$ is a free triple (15e and 15d). We have just shown that $y \geq_f x \geq_f y$ and $y \geq_f z \geq_f y$ both hold. And 1 dictates $f \mid \{x,z\}$ by Lemma 22 so $x >_f z >_f x$ holds. Therefore $f \mid \{x,y,z\}$ is a non-null social welfare function satisfying non-imposition. It is authoritarian by Theorem 5.4, and thus individual 1 must be a dictator for $f \,|\, \{x,y,z\}$. Then $y \in Y$ because x belongs to N. ∎

25. Lemma. $Y = X$ if T is finite.

Proof. Suppose that $X - Y \neq \varnothing$. Then $X - Y$ is not closed because Y is closed and X is connected. Choose any y belonging to Y and to the closure of $X - Y$. Now choose $x \in N \cap F(y)$ such that individual 1 dictates $f \,|\, \{x,y\}$. Choose $p, p' \in D$ such that $(x,y) \in Ap(1) \cap -Ap'(1)$. Then $(x,y) \in Af(p) \cap -Af(p')$. There exists a neighbourhood $N(y)$ such that $\{x\} \times N(y) \subset Af(p) \cap -Af(p')$. Now choose $z \in N(y) \cap (X - Y)$. We have $x >_f z >_f x$. By 15c there is some $v \in X$ such that $\{x,v\}$ and $\{z,v\}$ are free. By 15d there exist $w^1, w^2 \in N$ such that $\{w^1, w^2, z\}$ is free. Person 1 dictates $f \,|\, \{w^1, w^2\}$ by Lemma 22 and hence $w^1 >_f w^2 >_f w^1$. And $w^i >_f x >_f w^i$ for $i = 1,2$, by Lemma 22. Therefore, $w^i >_f x >_f z$ and $z >_f x >_f w^i$. Then $w^i >_f z >_f w^i$ by Lemma 17. Therefore $f \,|\, \{w^1, w^2, z\}$ is a non-null social welfare function satisfying non-imposition. Therefore, individual 1 must be a dictator for $f \,|\, \{w^1, w^2, z\}$ because 1 dictates $f \,|\, \{w^1, w^2\}$. This contradicts $z \notin Y$. Therefore, $X = Y$. ∎

26. Theorem. Assume that T is finite, X is a connected T_1-space, and D is standard. If $f : D \to P(X)$ satisfies Arrow's independence axiom then f is constant or authoritarian.

Proof. If f is not constant then by Lemmas 17–25 there exists an individual t such that either (i) t is a dictator for $f \,|\, \{x,y\}$ whenever $x \in X$ and $y \in X \cap F(x)$ or (ii) t is an inverse dictator for $f \,|\, \{x,y\}$ whenever $x \in X$ and $y \in X \cap F(x)$. (Note Lemmas 23 and 25.) Assume that (i) holds for $t = 1$. If $(x,y) \in Ap(1)$ then $x \in F_1(y)$ by property 15a. By 15c there exists $z \in X$ such that $\{x,z\}$ and $\{y,z\}$ are free. Then $D_1(\{x,y,z\}) = P(\{x,y,z\})$ by 15b. Then there is some $p' \in D$ such that $p'(t) = p(t) \,\forall\, t \neq 1$ and $(x,z), (z,y) \in Ap'(1)$. We have $(x,z), (z,y) \in Af(p')$ by (i). Therefore, $(x,y) \in Af(p')$ by transitivity. Thus $(x,y) \in Af(p)$ by Arrow's independence axiom. Case (ii) is handled in a similar fashion. ∎

We can go further and state that Arrow's independence axiom is equivalent to constancy or complete authoritarianism if D is standard and regular, and social preference is a continuous preorder in some connected T_1-topology (Proposition 10 and Theorem 26).

27. Theorem. Suppose that T is finite, X is a connected T_1-space, and D is standard and regular. Then $f: D \to P(X)$ satisfies Arrows independence axiom if and only if it is constant or completely authoritarian.

Proof. A constant or *completely* authoritarian social welfare function satisfies Arrow's independence axiom without restriction. And we have just pointed out that Proposition 10 and Theorem 26 establish the reverse inclusion under the hypothesis of this theorem. ∎

Full transitivity of social preference is indispensable for the preceding results. If we allow intransitivity of the social indifference relation then Arrow's independence axiom by itself does not imply oligarchy even with a continuity requirement. For instance . . .

28. Example. Choose two disjoint and non-empty open subsets Y and Z of X. (If X is a Euclidean space take two disjoint open spheres.) For *any* domain $D \subset P(X)^T$ let $f: D \to Q(X)$ be defined as follows:

$$f(p) \cap Y^2 = p(1) \cap Y^2.$$
$$f(p) \cap Z^2 = p(2) \cap Z^2.$$
$$(x,y) \in Sf(p) \text{ in all other situations.}$$

Clearly, f satisfies Arrow's independence axiom. To prove that $f(p)$ is continuous, choose $x \in X$ and let $X_1 = \{y \in X \mid (y,x) \in Af(p)\}$ and $X_2 = \{y \in X \mid (x,y) \in Af(p)\}$. If X_1 is empty it certainly is an open set. If $y \in X_1$ then $(y,x) \in Ap(t)$ for $t = 1$ or $t = 2$ and $x \in Y$ or $x \in Z$. Because $p(t)$ is continuous there is an open set $N(y)$ containing y such that $(z,x) \in Ap(t)$ for all $z \in N(y)$. Because Y and Z are open the sets $N(y) \cap Y$ and $N(y) \cap Z$ are also open. Therefore, either $(z,x) \in Af(p)$ for all $z \in N(y) \cap Y$ or $(z,x) \in Af(p)$ for all $z \in N(y) \cap Z$. In either case X_1 contains an open set $N'(y)$ that contains y. Therefore, X_1 is open as the union of open sets: $X_1 = \cup \{N'(y) \mid y \in X_1\}$. Similarly, X_2 is open. To prove that $f(p)$ is quasitransitive suppose we have $(x,y) \in Af(p)$ and $(y,z) \in Af(p)$. Then x, y, and z all belong to Y or they all belong to Z, and we have $(x,y),(y,z) \in Ap(t)$ for $t = 1$ or $t = 2$. Therefore $(x,z) \in Ap(t)$ because $Ap(t)$ is transitive. Therefore, $(x,z) \in Af(p)$.

The social welfare function of this example is not even quasi-oligarchical (3.24). By definition, individual 1 is a complete dictator for $f \,|\, Y$ and 2 is a complete dictator for $f \,|\, Z$. If the triple (I,J,K) is decisive over f itself then 1 and 2 must belong to I. But neither 1 nor 2 has veto power because the other is dictatorial over some subset of X. To prove an impossibility theorem for quasitransitive social preference in connected spaces we need to assume more than the Arrow independence axiom. We will add strict non-imposition to the hypothesis and derive four oligarchy theorems. The first result uses the free-triple property and regularity to prove the existence of a complete oligarchy. Completeness is a consequence of regularity, which is partly topological. Regularity also ensures that the quasi-oligarchy is actually an oligarchy. The next theorem assumes an unrestricted domain, $P(X)^{\mathrm{T}}$, and a normal and connected first-countable space to derive a complete oligarchy. Theorem 35 generalizes this by showing how strict non-imposition can be weakened to minimal responsiveness when the domain is $P(X)^{\mathrm{T}}$. (Note that Example 28 does not even satisfy minimal responsiveness: if $x \notin Y \cup Z$ then $(x,y) \in Sf(p)$ for all $y \in X$ and all $p \in D$.) The reader interested only in economic applications can skip to Theorem 36 which establishes oligarchy by assuming a standard domain.

29. Theorem. Let T be any finite set. Assume that D is a regular domain with the free-triple property and X has at least three members. A social welfare function $f: D \to Q(X)$ satisfies Arrow's independence axiom and strict non-imposition if and only if it is completely oligarchical.

Proof. (i) If f is completely oligarchical then there exist subsets I and J of T such that $Af(p) = \bigcap_{i \in I} Ap(i) \cap \bigcap_{j \in J} -Ap(j)$ for all $p \in D$. Continuity of $f(p)$ is easy to prove. If $(x,y) \in Af(p)$ then $(x,y) \in Ap(i)$ for all $i \in I$ and $(y,x) \in Ap(j)$ for all $j \in J$. By continuity of individual preference, for each $t \in T$ there is a neighbourhood $N_t(x)$ of x such that $(z,y) \in N_i(x)$ for all $i \in I$ and $(y,z) \in N_j(x)$ for all $j \in J$. Because T is finite the set $N(x) = \bigcap_{t \in I \cup J} N_t(x)$ is open. Clearly, $(z,y) \in \bigcap_{i \in I} Ap(i) \cap \bigcap_{j \in J} -Ap(j)$ for all $z \in N(x)$ and thus $\{z \in X \,|\, (z,y) \in Af(p)\}$ is open. Similarly, $\{z \in X \,|\, (x,z) \in Af(p)\}$ is open.

Because $Ap(t)$ and $-Ap(t)$ are transitive for each $t \in T$ the strict social preference relation $Af(p)$ is transitive. If x and y are distinct then by the free-triple property there is some $p \in D$ such that $(x,y) \in Ap(i) \cap -Ap(j)$ for all $i \in I$ and $j \in J$. Hence $(x,y) \in$

$Af(p)$ for some $p \in D$ and f satisfies strict non-imposition. It obviously satisfies Arrow's independence axiom.

(ii) Suppose that $f: D \to Q(X)$ satisfies strict non-imposition and Arrow's independence axiom. Then f is quasi-oligarchical by Theorem 6.3. Let (I,J,K) denote the quasi-oligarchy for f. Suppose we have $t \in K$. By the free-triple property we can find distinct x and y in X such that

$$(x,y) \in \cap_{i \in I} Ap(i) \cap_{j \in J} - Ap(j) \cap_{k \in K} Sp(k)$$

and hence $(x,y) \in Af(p)$. By regularity we can find $p' \in D$ such that $(x,y) \in Sp'(t), p'(m) = p(m) \; \forall \, m \neq t$, and every neighbourhood of x contains a point z such that $(z,x) \in Ap'(t)$ and thus $(z,y) \in Ap'(t)$. We have $(x,y) \in Af(p')$ by the independence axiom. Because the members of K have quasi-veto power, every neighbourhood of x contains a point z such that $(z,y) \notin Af(p')$, contradicting continuity of $f(p')$. Therefore, $K = \emptyset$ and f is oligarchical. And f is completely oligarchical by Proposition 10. ∎

Lemmas 12 and 13 along with Theorem 29 immediately yield the following.

30. Theorem. Let T be a finite set and let X be a normal and connected T_1-space satisfying the first axiom of countability. Then $f: P(X)^T \to Q(X)$ satisfies Arrow's independence axiom and strict non-imposition if and only if it is completely oligarchical.

We can sharpen Theorem 30 by substituting for strict non-imposition the much weaker minimal responsiveness condition (3.4). Four new lemmas are required. Recall that $x >_f y$ means that (x,y) belongs to $Af(p)$ for some $p \in D$ (4.1). Now define the *strict component* $H(x)$ of the social welfare function f:

$$H(x) = \{x\} \cup \{y \in X \mid x >_f y \text{ and } y >_f x\}.$$

31. Lemma. Let T be any set. If D has the free-triple property and $f: D \to Q(X)$ satisfies Arrow's independence axiom then for any three distinct $x,y,z \in X$ we have $x >_f z$ if $x >_f y >_f z$.

Proof. Assume that $(x,y) \in Af(p)$, $(y,z) \in Af(p')$ and $x \neq y \neq z \neq x$. The free-triple property implies that there is some $p'' \in D$ such that $p''(t) \cap \{x,y\}^2 = p(t) \cap \{x,y\}^2$ and $p''(t) \cap \{y,z\}^2 = p'(t) \cap \{y,z\}^2$ for all $t \in T$. Then $(x,y),(y,z) \in Af(p'')$ by the independence axiom and $(x,z) \in Af(p'')$ by transitivity of $Af(p'')$. Therefore, $x >_f z$. ∎

32. Lemma. Let T be any set. If D has the free-triple property and $f: D \to Q(X)$ satisfies Arrow's independence axiom then $y \in H(x)$ implies $H(x) = H(y)$.

Proof. Certainly $H(x) = H(y)$ if $x = y$. Assume that $x \neq y$ and $y \in H(x)$. We have $x \in H(y)$ by definition of $H(x)$ and the fact that y belongs to $H(x)$. If $x \neq z \neq y$ and $x >_f z >_f x$ then $y >_f x >_f z >_f x >_f y$ and thus $y >_f z >_f y$ by Lemma 31. Therefore, $z \in H(y)$. Similarly, $x \neq z \neq y$ and $z \in H(y)$ implies $z \in H(x)$. Therefore, $H(x) = H(y)$. ∎

33. Lemma. Let T be any set. If D has the free-triple property and $f: D \to Q(X)$ satisfies Arrow's independence axiom then $H(x) \cap H(y) \neq \varnothing$ implies $H(x) = H(y)$.

Proof. If $z \in H(x) \cap H(y)$ then $H(x) = H(z) = H(y)$ by Lemma 32. ∎

34. Lemma. Let T be any set. If D has the free-triple property and $f: D \to Q(X)$ satisfies Arrow's independence axiom and minimal responsiveness then $H(x)$ is open.

Proof. By definition, $H(x) = \{y \in X \mid x >_f y >_f x\} \cup \{x\}$. $H(x) \neq \{x\}$ by minimal responsiveness. Choose $z \in H(x)$ distinct from x. Then $(x,z) \in Af(p)$ and $(z,x) \in Af(p')$ for appropriate choices of p and p'. By continuity of $f(p)$ and $f(p')$ there are neighbourhoods $N(x)$ and $N'(x)$ of x and neighbourhoods $N(z)$ and $N'(z)$ of z such that $(v,w) \in Af(p)$ if $v \in N(x)$, $w \in N(z)$, and $(w,v) \in Af(p')$ if $w \in N'(z)$, $v \in N'(x)$. Setting $v = x$ gives us $N(z) \cap N'(z) \subset H(x)$. Setting $w = z$ yields $N(x) \cap N'(x) \subset H(z)$. Because $H(x) = H(z)$ by Lemma 33 we also have $N(x) \cap N'(x) \subset H(x)$. Therefore, every member of $H(x)$ is contained in some open subset of $H(x)$. Therefore, $H(x)$ is open. ∎

Now we can prove a much stronger version of Theorem 30.

35. Theorem. Let T be a finite set and let X be a normal and connected T_1-space satisfying the first axiom of countability. Then $f: P(X)^T \to Q(X)$ satisfies Arrow's independence axiom and minimal responsiveness if and only if it is completely oligarchical.

Proof. (i) By Theorem 29 a completely oligarchical social welfare function satisfies the independence axiom and strict non-imposition. Because strict non-imposition implies minimal responsiveness we have the 'if' part of our theorem.

(ii) Suppose f satisfies minimal responsiveness and Arrow's independence axiom. Choose any $x \in X$. $H(x)$ is open by Lemma 34. Suppose $H(y) \neq H(x)$ for some $y \in X$. Then $Y = \cup \{H(y) \mid y \in X \text{ and } H(y) \neq H(x)\}$ is well defined and non-empty. It is open as the union of open sets (Lemma 34). $H(x) \cap Y$ is empty by Lemma 33. Of course $x \in H(x)$. Therefore, X is the union of two non-empty and disjoint open sets, contradicting connectedness. Therefore, $H(x) = H(y)$ for all $y \in X$. Because $y \in H(y)$, and hence $\cup_{y \in X} H(y) = X$, we must have $H(x) = X$. Therefore, f satisfies strict non-imposition: if $y \neq z$ then $z \in H(y)$ and hence $z >_f y >_f z$. Therefore, f is completely oligarchical by Theorem 30. ■

Now we turn our attention to a result that will be a stepping-stone to the theorems in Chapters 10−13 on economic environments. The results on transitive social preference are based on Arrow's independence axion alone; a non-imposition condition is not required because of the power of transitivity of social indifference. In the case of quasitransitive social preference and oligopoly, the strict non-imposition axiom carries much of the burden of proof.

36. Theorem. Assume that T is a finite set and D is standard and regular. A social welfare function $f \colon D \to Q(X)$ satisfies Arrow's independence axiom and strict non-imposition if and only if it is completely oligarchical.

Proof. (i) If f is completely oligarchical then it satisfies Arrow's independence axiom and strict non-imposition by Theorem 29.

(ii) Suppose that $f \colon D \to Q(X)$ satisfies strict non-imposition and Arrow's independence axiom. There exists a free triple $Z = \{x^*, y^*, z^*\}$ by 15c and 15d. Then $f \mid Z$ is quasi-oligarchical by Theorem 6.3. Let (I, J, K) be the oligarchy underlying $f \mid Z$. Suppose that $\{x, y\}$ is any free pair. Then there exist $w, z \in X$ such that $Z^1 = \{x^*, y^*, w\}$, $Z^2 = \{y^*, w, z\}$, $Z^3 = \{w, z, x\}$, and $Z^4 = \{z, x, y\}$ are each free triples. Then each $f \mid Z^i$ is quasi-oligarchical by Theorem 6.3. Because $Z^i \cap Z^{i+1}$ has a free pair in common and an oligarchy possesses both decisiveness *and* veto power, the oligarchies must be the same for $f \mid Z^i$ and $f \mid Z^{i+1}$. Then we have proved the following: if (I, J, K) is the quasi-oligarchy underlying $f \mid Z$ then for any free pair $\{x, y\}$ the social welfare function $f \mid \{x, y\}$ is quasi-oligarchical and (I, J, K) is the underlying quasi-oligarchy.

Suppose that $(x, y) \in Ap(i) \cap -Ap(j) \cap Sp(k)$ for all $i \in I$,

$j \in J$, and $k \in K$. By 15c we can choose $z \in X$ such that $\{x,z\}$ and $\{y,z\}$ are both free. Now choose $p' \in D$ such that $p'(t) = p(t)$ for all $t \notin I \cup J \cup K$ and (x,z), $(z,y) \in Ap'(i) \cap -Ap'(j) \cap Sp(k)$ for all $i \in I$, $j \in J$, and $k \in K$. This is possible by 15a and 15b: if $(x,y) \in Ap(t)$ then $x \in F_t(y)$ by 15a and thus $D_t(\{x,y,z\}) = P(\{x,y,z\})$ by 15b. If $(x,y) \in Sp(t)$ and $x \notin F_t(y)$ then $(x,y) \in SR \, \forall \, R \in D_t$ by 15a. Because $x \in F_t(z)$ we have $(x,z) \in SR'$ for some $R' \in D_t$ and thus $(z,y) \in SR'$ because R' is complete and transitive. Now we have (x,z), $(z,y) \in Af(p')$ because (I,J,K) is a quasi-oligarchy for $f \mid \{x,z\}$ and $f \mid \{z,y\}$. Then $(x,y) \in Af(p')$ by transitivity and $(x,y) \in Af(p)$ by Arrow's independence axiom. Therefore, (I,J,K) is decisive for all ordered pairs of outcomes.

Suppose $i \in I$ and $(x,y) \in Ap(i) \cap -Af(p)$. There exist neighbourhoods $N(x)$ and $N(y)$ such that $N(x) \times N(y) \subset Ap(i) \cap -Af(p)$. Choose $x' \in N(x)$ and $y' \in N(y)$ such that $\{x',y'\}$ is free (15e). Then (I,J,K) is a quasi-oligarchy for $f \mid \{x',y'\}$ and hence i has veto power for $f \mid \{x',y'\}$, contradicting $(x',y') \in Ap(i) \cap -Af(p)$. Therefore, every member of I has veto power for f. Similarly, every member of J has inverse veto power for f. Suppose $k \in K$ and $(x,y) \in Ap(k) \cap -Af(p)$. We are again led to a contradiction because k has quasi-veto power for $f \mid \{x',y'\}$ if $\{x',y'\}$ is free. Similarly, $(x,y) \in Ap(k) \cap Af(p)$ is impossible if k belongs to K. Therefore, f is quasi-oligarchical. It is in fact oligarchical by the argument of part (ii) of Theorem 29. That is, K is empty. And f is completely oligarchical by Proposition 10. ∎

9 Infinite Societies

FISHBURN (1970) first demonstrated that for any infinite society T there exists a non-dictatorial social welfare function $f: P(X)^T \to P(X)$ satisfying the Pareto criterion and Arrow's independence axiom if $P(X)$ is the set of all preorders on X; that is, if the discrete topology is assumed. The proof is based on Zorn's lemma (1.36) which is used to establish the existence of a free ultrafilter (Proposition 5.13). It is not surprising that any social choice formula based on a free ultrafilter is incompatible with continuity of social preference in a sensible topology, as we demonstrate in this chapter. We extend the impossibility theorems of Arrow and Wilson to the case of a countable society and an outcome space that is a connected and first countable T_1-space. In plain words the impossibility theorems go through when — but only when — we require continuity of social preference with respect to a reasonably demanding topology. It is also necessary to assume that the domain is complete: any preorder on any finite subset of X is contained in some member of D_t (3.20). For application to economic environments completeness is too strong so we present other versions of the theorem that do translate to a standard economic framework. The oligarchy theorems of Chapter 6 are also extended to the infinite society case. We begin with a new domain assumption.

1. Basic domain. Assume that T is the set of positive integers. Then D is basic if it is standard (8.15) and for all distinct $x, y \in X$ there exist sequences $\{x^n\}$ and $\{y^n\}$ converging to x and y respectively, and a profile $p \in D$ such that $(y^n, x^n) \in Ap(t)$ for all $t \in T$ and $n \le t$, and $(x, y) \in Ap(t)$ for all $t \in T$ such that $x \in F_t(y)$.

Consider again the Cobb–Douglas space $Y = \Pi_{t \in T} Y_t$ endowed with the product topology. $Y_t = \{a \in \mathbf{E}^2_{++} \mid \log a_1 + \log a_2 = 0\}$. If $D = W(Y)$ then D is basic. (W is the domain of classical economic preferences over the entire allocation space Ω.) It is easy to see that $W(Y)$ is standard. To verify the second part of the definition note that for any two allocations $x, y \in Y$ and for any individual t we can easily find sequences $\{x^n(t)\}$ and $\{y^n(t)\}$ in arbitrary neighbour-hoods of $x(t)$ and $y(t)$ respectively so that $x_1^n(t) > x_1^{n+1}(t) > x_1(t) \geq y_1(t) > y_1^{n+1}(t) > y_1^n$ for all n, or else the reverse inequalities hold for all n, and $\{x^n\} \cap \{y^n\} = \varnothing$. Lemma 12.1 proves that the necessary preorders on these sequences can be generated by economic preferences, as one can easily verify by experimenting with a diagram. Note that this argument works with many topologies that are weaker than the product topology. One simply needs to verify that the topology is consistent with the existence of the desired sequences and with definition 8.15.

If $D = P(X)^T$ and X is a normal and connected T_1-space satisfy-ing the first axiom of countability then D has the free-triple property by Lemma 8.12 and the fact that a finite subset of X is a G_δ set (8.11). The existence of the sequences is guaranteed by Proposition 2.8 and the existence of the desired profile is guaranteed by Lemma 8.12. Therefore, $P(X)^T$ is a basic domain for any set T.

The first result extends Lemma 5.12 on the algebraic structure of decisive and inversely decisive coalitions to the family of standard domains. Recall that U_f is the family of decisive coalitions with respect to f, and V_f is the family of inversely decisive coalitions.

2. Lemma. Let T be any set. Assume that X is a connected T_1-space and D is standard. If $f\colon D \to P(X)$ is a non-null social welfare function satisfying non-imposition and Arrow's indepen-dence axiom then either U_f or V_f is an ultrafilter.

Proof. If f is not null then X is not a singleton set. Recall that an open set in a connected T_1-space cannot be finite unless the space has only one point (Lemma 8.7). Then properties 15c and 15d of 8.15 imply that for every $x \in X$ every open set contains two points y and z such that $\{x,y,z\}$ is a free triple. And because X is a T_1-space every finite subset Y is a discrete subspace and hence $P(Y)$ is the set of all preorders on Y.

Now, suppose that f is constant. Because f is not null (3.25) we have $(x,y) \in Af(p)$ for some $x,y \in X$ and $p \in D$. By continuity of $f(p)$ there exist neighbourhoods $N(x)$ and $N(y)$ of x and y respec-tively such that $(x',y') \in Af(p)$ for all $x' \in N(x), y \in N(y)$. Choose x' and y' such that $\{x',y'\}$ is free (15e of 18.15). Then $(x',y') \in Af(p')$ for all $p' \in D$ because f is constant, contradict-ing non-imposition. Therefore, f cannot be constant. Then by Lemmas 8.17–8.19 there is a three-element subset Z of X such that

$D(Z) = P(Z)^T$ and $f|Z$ is a non-null social welfare function satisfying non-imposition. Therefore, either $U_{f|Z}$ or $V_{f|Z}$ is an ultrafilter by Lemma 5.12. Set $U^* = U_{f|Z}$ and assume that it is an ultrafilter. Let $Z = \{x^*, y^*, z^*\}$.

Let $\{x,y\}$ be any free pair. Because D is standard there exist $w, z \in X$ such that $B = \{x^*, y^*, w\}$, $C = \{y^*, w, z\}$, $E = \{w, z, x\}$, and $F = \{z, x, y\}$ are free triples (15d of 18.15). We have $x^* >_f y^* >_f x^*$ because $U^* \neq \varnothing$ and $D(Z) = P(Z)^T$. Therefore, $f|B$ is a non-null social welfare function satisfying non-imposition. Therefore, either $U_{f|B}$ or $V_{f|B}$ is an ultrafilter, by Lemma 5.12. Because $B \cap Z$ has two members and $D(B \cap Z) = P(B \cap Z)^T$ we must have $U^* = U_{f|B}$ by Lemmas 4.3 and 4.4. Similarly, $U_{f|B} = U_{f|C} = U_{f|E} = U_{f|F}$. Therefore, every member of U^* is decisive for all (x,y) such that $\{x,y\}$ is free. (Even though we are considering different social welfare functions, f, $f|Z$, $f|B$, etc., the term decisive is unambiguous by virtue of the independence axiom.) $I \notin U^*$ cannot be decisive over (x,y) if $\{x,y\}$ is free. This follows from the fact that $T - I$ belongs to U^* if I does not, and hence $T - I$ is decisive for (x,y). Now suppose $p \in D$ and $(x,y) \in \bigcap_{t \in I} Ap(t)$ for arbitrary $x, y \in X$ and $I \in U^*$. By properties 15a, 15b, and 15c of 18.15 there is some $z \in X$ such that $\{x,z\}$ and $\{z,y\}$ are both free and $D_t(\{x,y,z\}) = P(\{x,y,z\})$ for all $t \in I$. Now choose $p' \in D$ such that $p'(t) = p(t)$ for all $t \notin I$ and $(x,z), (z,y) \in Ap'(t)$ for all $t \in I$. We have $(x,z), (z,y) \in Af(p')$ because I is decisive for all free pairs. Then $(x,y) \in Af(p')$ by transitivity of $Af(p')$ and thus $(x,y) \in Af(p)$ by Arrow's independence axiom. And, if $I \notin U^*$ then $T - 1$ is decisive for free pairs so I is not decisive. Therefore $U_f = U^*$, an ultrafilter. Similarly, V_f is an ultrafilter if $V_{f|Z}$ is. ∎

3. Theorem. Assume that T is countable, X is a connected and first-countable T_1-space, and D is complete. If $f: D \rightarrow P(X)^T$ is a non-null social welfare function satisfying non-imposition and Arrow's independence axiom then f is authoritarian.

Proof. If X is a singleton the theorem is true vacuously. If X is not a singleton it must be infinite (Proposition 2.5). Completeness of D (3.20) obviously implies that D is standard if the connected space X is T_1. Therefore, either U_f or V_f is an ultrafilter, by Lemma 2. Suppose that U_f is an ultrafilter.

If U_f is fixed then there is some $t \in T$ such that $U_f =$

$\{I \subset T \mid t \in I\}$ and t is a dictator. Suppose that U_f is free. That is, $\cap U_f = \varnothing$.

Without loss of generality assume that T is the set of positive integers. Choose distinct $x, y \in X$. By Proposition 2.8 there exist sequences $\{x^i\}$ and $\{b^i\}$ in $X - \{x\}$ and $X - \{y\}$ respectively such that $\{x^i\}$ converges to x, $\{b^i\}$ converges to y, and $x^i \neq x^j \neq x$, $b^i \neq b^j \neq y$ if $i \neq j$. Because the space is T_1 there exist open sets $N(y)$ and $N_i(y)$, $\forall i \in T$, such that $x \notin N(y)$ and $x^i \notin N_i(y)$. Set $M_n(y) = N(y) \cap N_1(y) \cap N_2(y) \ldots \cap N_n(y)$. Because $\{b^n\}$ converges to y, for each $t \in T$ there is some $i \geq t$ such that $y^i \in M_n(y)$. Set $y^i = b^i$. Then $\{y^i\}$ converges to y, $x^i \neq x \neq y^i$, and $x^i \neq y^j$ for $j \geq i$.

Because D is complete (and a product set) there is some $p \in D$ such that for each $t \in T$ we have $(x, y) \in Ap(t)$ and $(y^n, x^n) \in Ap(t)$ for all $n \leq t$. We may have $y^i = x^j$ for $j < i \leq t$, but never for $j \geq t$, so transitivity of $p(t)$ will not be sacrificed. T belongs to U_f because U_f is an ultrafilter. Because $\cap U_f = \varnothing$, for each $t \in T$ there is some $I_t \in U_f$ such that $t \notin I_t$. Then $I_1 \cap I_2 \cap \ldots \cap I_t$ belongs to U_f and therefore so does any superset; the set $T_t = T - \{1, 2, \ldots, t\}$, in particular. We have $(x, y) \in Af(p)$ because T belongs to U_f. But for any $t \in T$ we have $(y^t, x^t) \in Ap(n)$ for all $n \in T_{t-1}$ and thus $(y^t, x^t) \in Af(p)$ for all $t \in T$. Therefore, $f(p)$ cannot be continuous. Therefore, U_f is a fixed ultrafilter if it is an ultrafilter and $f: D \to P(X)$ is a non-null social welfare function satisfying non-imposition and Arrow's independence axiom.

The obvious modifications of the above argument establish that V_f is fixed if it is an ultrafilter, in which case there is an inverse dictator. ∎

The critical sequences $\{x^n\}$ and $\{y^n\}$ and profile p employed in the proof of Theorem 3 are incorporated in the definition of a basic domain. Therefore, we have already proved the next result.

4. Theorem. Assume that T is countable, X is a connected T_1-space, and D is a basic domain. If $f: D \to P(X)$ is a non-null social welfare function satisfying non-imposition and Arrow's independence axiom then f is authoritarian.

Theorem 4 appears to rely less on the topological apparatus than does Theorem 3, but that is illusory. The definition of a basic domain makes demands on the topology of the outcome space.

To enable non-constancy to be substituted for the assumption that f is non-null and non-imposed it is necessary to assume that X is *locally* connected. A space is locally connected if for every point x and every open set N containing x there is a connected and open subset N' of X such that $x \in N' \subset N$. (See Dugundji 1966: 113.)

5. Theorem. Assume that T is countable and X is a first-countable T_1-space that is both connected and locally connected. If D is complete and $f: D \to P(X)$ satisfies Arrow's independence axiom then f is either constant or authoritarian.

Proof. The theorem is obviously true if X is a singleton. Therefore, we may assume that X is not a singleton. If f is not constant then by Lemma 8.6 there is some component G of f such that $f \mid G$ is not constant and G contains an open set N. Then we may assume that N is connected because X is locally connected. N is infinite as an open subset of a connected T_1-space (Lemma 8.7). Suppose that $f \mid N$ is constant. Because $f \mid G$ satisfies non-imposition by Lemma 8.1 and $f \mid G$ is not constant, either $U_{f \mid G}$ or $V_{f \mid G}$ is an ultrafilter by Lemma 5.12. Because N is infinite and D has the free-triple property, constancy of $f \mid N$ implies that $U_{f \mid N}$ and $V_{f \mid N}$ are both empty. Because $N \subset G$ we have a contradiction: either $U_{f \mid G}$ or $V_{f \mid G}$ is an ultrafilter. Therefore, $f \mid N$ is not constant.

Now, $f \mid N$ is a non-null social welfare function mapping $D(N)$ into $P(N)$. N is a connected and first-countable T_1-space, $D(N)$ is complete, and $f \mid N$ satisfies non-imposition and Arrow's independence axiom. Therefore, $f \mid N$ is authoritarian by Theorem 3. But $U_f = U_{f \mid N}$ and $V_f = V_{f \mid N}$ by Lemmas 4.3 and 4.4. Therefore, f is authoritarian. ∎

6. Theorem. Assume that T is countable and X is a connected T_1-space that is also locally connected. If D is a regular basic domain and $f: D \to P(X)$ is a non-constant social welfare function satisfying Arrow's independence axiom then f is completely authoritarian.

Proof. By Lemmas 8.17–8.19 there exist disjoint and non-empty open sets N and M such that $x >_f y >_f x$ for all $x \in N$, $y \in M$. Because X is locally connected we can assume that N is connected. Choose a three-element set $Z = \{x^*, y^*, z^*\}$ such that Z is free, $x^*, y^* \in N$, and $z^* \in M$. Then $f \mid Z$ is a non-null welfare function satisfying non-imposition (Lemma 8.19). Therefore, either $U^* \equiv U_{f \mid Z}$ or $V^* \equiv V_{f \mid Z}$ is an ultrafilter by Lemma 5.12. Suppose that U^* is an ultrafilter. If x and y belong to N and $\{x,y\}$ is free then choose distinct $w,z \in M$ such that $B = \{x^*,y^*,w\}$, $C = \{y^*,w,z\}$, $E = \{w,z,x\}$, and $F = \{z,x,y\}$ are free triples. Then $f\,|\,B$, $f\,|\,C$, $f\,|\,E$, and $f\,|\,F$ are each non-null social welfare functions satisfying non-imposition by Lemma 8.19. To prove this note that $x^* >_f y^* >_f x^*$ because $U^* \neq \varnothing$ and $D(Z) = P(Z)^{\mathrm{T}}$. Therefore, $f\,|\,B$ is a non-null social welfare function satisfying non-imposition. Therefore, either $U_{f|\mathrm{B}}$ or $V_{f|\mathrm{B}}$ is an ultrafilter, by Lemma 5.12. Because $B \cap Z$ has two members and $D(B \cap Z) = P(B \cap Z)^{\mathrm{T}}$ we must have $U^* = U_{f|\mathrm{B}}$ by Lemmas 4.3 and 4.4. Similarly, $U_{f|\mathrm{B}} = U_{f|\mathrm{C}} = U_{f|\mathrm{E}} = U_{f|\mathrm{F}}$. Therefore, every member of U^* is decisive for all (x,y) such that $\{x,y\} \subset N$ and $\{x,y\}$ is free. $I \notin U^*$ cannot be decisive over (x,y) if $\{x,y\}$ is free. This follows from the fact that $T - I$ belongs to U^* if I does not, and hence $T - I$ is decisive for (x,y). Now suppose $p \in D$ and $(x,y) \in \cap_{t \in I} Ap(t)$ for arbitrary $x,y \in N$ and $I \in U^*$. By properties 15a, 15b, and 15e of 18.15 there is some $z \in N$ such that $\{x,z\}$ and $\{z,y\}$ are both free and $D_t(\{x,y,z\}) = P(\{x,y,z\})$ for all $t \in I$. Now choose $p' \in D$ such that $p'(t) = p(t)$ for all $t \notin I$ and $(x,z),(z,y) \in Ap'(t)$ for all $t \in I$. We have $(x,z),(z,y) \in Af(p')$ because I is decisive for all free pairs in N. Then $(x,y) \in Af(p')$ by transitivity of $Af(p')$ and thus $(x,y) \in Af(p)$ by Arrow's independence axiom. And, if $I \notin U^*$ then $T - I$ is decisive for free pairs so I is not decisive. Therefore $U_{f|\mathrm{N}} = U^*$, an ultrafilter. Obviously $f\,|\,N$ is not constant. Now we show that $f\,|\,N$ is authoritarian.

We know that $x >_f y$ holds if $x,y, \in N$ and $(x,y) \in \cap_{t \in I} Ap(t)$ for some $I \in U^*$ and $p \in D$. Suppose that x and y belong to N and $(y,x) \in Af(p)$ for all $p \in D$. Then $y \neq x$. Choose sequences $\{x^n\}$ and $\{y^n\}$ converging to x and y respectively and profile $p \in D$ such that $(x^n,y^n) \in Ap(t)$ for all $t \in T$ and all $n \leq t$. We may assume that $\{x^n\}$ and $\{y^n\}$ are subsets of N. We have $(y,x) \in Af(p)$. By continuity there exist neighbourhoods $N(y)$ and $N(x)$ such that $N(y) \times N(x) \subset Af(p)$. Then there is some t^* such that $(y^n,x^n) \in Af(p)$ for all $n \geq t^*$. Then $T^* = T - \{1,2,\ldots,t^* - 1\} \notin U^*$. Therefore, $I = \{1,2,\ldots,t^* - 1\}$ belongs to U^*. If $\{n\} \notin U^*$ and $n \leq t^* - 1$ then $T - \{n\} \in U^*$ and thus $I - \{n\} = (T - \{n\}) \cap I$ belongs to U^*. Therefore, if $\{n\} \notin U^*$ for all $n < t^* - 1$ we have $(I - \{1\}) \cap (I - \{2\}) \cap \ldots \cap (I - \{t^* - 2\}) = \{t^* - 1\} \in U^*$. Therefore, U^* is fixed. Then $f\,|\,N$ is dictatorial and hence is completely dictatorial by Proposition 8.10. Therefore, $f\,|\,N$

satisfies non-imposition by 15a of 8.15: if t is the dictator and $D_t(\{x,y\}) \neq P(\{x,y\})$ then $(x,y) \in Sp(t)$ for all p and thus $(x,y) \in Sf(p)$ for all p.

We know that $f|N$ is a non-null social welfare function satisfying non-imposition and Arrow's independence axiom. Because N is a connected T_1-space and $D(N)$ is basic the social welfare function $f|N$ is authoritarian by Theorem 4. Suppose that $f|N$ is dictatorial and that 1 is the dictator. Let Y denote the set of $x \in X$ such that 1 is a dictator for $f|\{x,y\}$ for some $y \in N \cap F(x)$.

Claim 1. If $x \in Y$ and $y \in Y \cap F(x)$ then 1 is a dictator for $f|\{x,y\}$.

Proof of claim 1. If x and y belong to Y and $x \neq y$ then there exist x' and y' in N such that 1 dictates $f|\{x,x'\}$ and $f|\{y,y'\}$ and both $\{x,x'\}$ and $\{y,y'\}$ are free. Then we have $x >_f x' >_f x$ and $y >_f y' >_f y$. Therefore there exist open subsets $N(x')$ and $N(y')$ of N containing x' and y' respectively and such that $x >_f v >_f x$ and $y >_f w >_f y$ for all $v \in N(x')$, $w \in N(y')$. Choose v and w so that $\{x,y,v\}$ and $\{y,v,w\}$ are free triples (15d). Then 1 dictates $f|\{v,w\}$. Therefore, $v >_f w >_f v$. Now we have $x >_f v >_f w >_f y >_f w >_f v >_f x$ and hence both $f|\{y,v,w\}$ and $f|\{x,y,v\}$ satisfy non-imposition by Lemma 8.17. Therefore, 1 dictates $f|\{y,v,w\}$ by Lemma 5.12 and the fact that 1 dictates $f|\{v,w\}$. Thus 1 dictates $f|\{x,y,v\}$ by Lemma 5.12 and the fact that 1 dictates $f|\{y,v\}$. This proves claim 1.

Claim 2. Y is closed.

Proof of claim 2. Suppose that y belongs to the closure of Y. Choose $x \in N \cap F(y)$. (The set is not empty by 15e.) Suppose $(y,x) \in Af(p)$ for all $p \in D$. Choose $p \in D$ such that $(x,y) \in Ap(1)$. We have $(y,x) \in Af(p)$. There exist neighbourhoods $N(x)$ and $N(y)$ such that $N(y) \times N(x) \subset [Af(p) \cap -Ap(1)]$. Choose $y' \in N(y) \cap Y$. By 15c there is some $y'' \in X$ such that $\{y',y''\}$ is free and hence by 15d we can find some $x' \in N(x) \cap N \cap F(y')$. We have $x' \in Y$ by 15e and the fact that 1 dictates $f|N$. Then 1 dictates $f|\{x',y'\}$ by claim 1. Then $(x',y') \in Af(p)$, a contradiction. Similarly, we can rule out the possibility that $x \in N \cap F(y)$ and $(x,y) \in Af(p)$ holds for all $p \in D$. Now choose $x,z \in N$ such that $\{x,y,z\}$ is free (15e and 15d). We have just shown that $y \geq_f x \geq_f y$ and $y \geq_f z \geq_f y$ both hold. And 1 dictates $f|\{x,z\}$ so $x >_f z >_f x$ holds. Therefore $f|\{x,y,z\}$ is a non-null social welfare function satisfying non-imposition. Thus individual 1 must be a dictator for $f \mid \{x,y,z\}$ by Lemma 5.12 and the fact that 1 dictates $f \mid \{x,z\}$. Then $y \in Y$ because x belongs to N, and we have proved claim 2.

Claim 3. $Y = X$.

Proof of claim 3. Suppose that $X - Y \neq \varnothing$. Then $X - Y$ is not closed because Y is closed and X is connected. Choose any y belonging to Y and to the closure of $X - Y$. Now choose $x \in N \cap F(y)$ such that 1 dictates $f \mid \{x,y\}$. Choose $p,p' \in D$ such that $(x,y) \in Ap(1) \cap -Ap'(1)$. Then $(x,y) \in Af(p) \cap -Af(p')$. There exists a neighbourhood $N(y)$ such that $\{x\} \times N(y) \subset Af(p) \cap -Af(p')$. Now choose $z \in N(y) \cap (X - Y)$. We have $x >_f z >_f x$. By 15c there is some $v \in X$ such that $\{x,v\}$ and $\{z,v\}$ are free. By 15d there exist w^1, $w^2 \in N$ such that $\{x,w^1,w^2\}$ and $\{w^1,w^2,z\}$ are free triples. Person 1 dictates $f \mid \{w^1,w^2\}$ and hence $w^1 >_f w^2 >_f w^1$. And $w^i >_f x >_f w^i$ for $i = 1,2$ because w^i and x belong to N. Therefore, $w^i >_f x >_f z$ and $z >_f x >_f w^i$. Then $w^i >_f z >_f w^i$ by Lemma 8.17. Therefore $f \mid \{w^1,w^2,z\}$ is a non-null social welfare function satisfying non-imposition. Therefore, 1 must be a dictator for $f \mid \{w^1,w^2,z\}$ by Lemma 5.12 because 1 dictates $f \mid \{w^1,w^2\}$. This contradicts $z \notin Y$. Therefore, $X = Y$.

We know that 1 is a dictator for $f \mid \{x,y\}$ whenever $x \in X$ and $y \in X \cap F(x)$ (Claims 1 and 3). Let x and y be arbitrary members of X. If $(x,y) \in Ap(1)$ then $x \in F_1(y)$ by property 15a of 18.15. By 15c there exists $z \in X$ such that $\{x,z\}$ and $\{y,z\}$ are free. Then $D_1(\{x,y,z\}) = P(\{x,y,z\})$ by 15b. Then there is some $p' \in D$ such that $p'(t) = p(t) \; \forall \, t \neq 1$ and $(x,z),(z,y) \in Ap'(1)$. We have $(x,z),(z,y) \in Af(p')$ because $\{x,z\}$ and $\{z,y\}$ are both free. Therefore, $(x,y) \in Af(p')$ by transitivity. Thus $(x,y) \in Af(p)$ by Arrow's independence axiom. Therefore, f is dictatorial. If we begin by assuming that V^* is an ultrafilter then a parallel argument shows that f is inversely dictatorial. Proposition 8.10 completes the proof. ■

If for no other reason than comparison with the original theorems of Arrow and Wilson we present an 'unrestricted domain' version of the infinite society impossibility theorem.

7. Theorem. Assume that T is countable and that X is a normal and connected first-countable T_1-space that is also locally connected.

Then $f: P(X)^T \to P(X)$ satisfies Arrow's independence axiom if and only if it is constant or completely authoritarian.

Proof. (i) A constant or completely authoritarian social welfare function satisfies Arrow's independence axiom without restriction.

(ii) $P(X)^T$ is complete by Lemmas 8.11 and 8.12. Hence it is standard. It is regular by Lemma 8.13. By Proposition 2.8 the domain is basic. (See the argument of Theorem 3.) Therefore, a social welfare function $f: P(X)^T \to P(X)$ satisfying Arrow's independence axiom is constant or completely authoritarian by Theorem 6. ∎

Assuming that T is an atomless measure space, Kirman and Sondermann (1972) prove that there exists a decisive coalition of arbitrarily small measure if f maps $P(X)$ into the set of *all* preorders on X and satisfies the Pareto criterion and Arrow's independence axiom. A subsequent topological argument justifies the claim that there is an 'invisible' dictator. We have been able to show that there is a virtual dictator (or inverse dictator) if social preference is required to be continuous, and this conclusion obtains even if the Pareto criterion is weakened to the mildest possible responsiveness condition, non-constancy of the social welfare function.

Now we turn to the case of quasitransitive social preference. Extension of the oligarchy theorem to a countable T will require a new domain assumption. Transitivity of social indifference bears much of the burden of proof when full transitivity of social preference is assumed. When mere quasitransitivity is imposed we must assume a more complex domain. Let X be a topological space and let the product set D be a subset of $P(X)^T$.

8. Elementary domain. The domain D is elementary if it is standard and there exists a free pair $\{x,y\}$ and sequences $\{x^i\}$ and $\{y^i\}$ converging to x and y respectively such that

8a $x^i \neq x^j \neq x$, $y^i \neq y^j \neq y$ if $i \neq j$.

8b $\{x^n, y^n\}$ is free for all n.

8c For each $t \in T$ there exist preorders $R^1(t)$, $R^2(t)$, $R^3(t)$, $R^4(t)$ and $R^5(t)$ in D_t with the following properties:

$(x,y^n), (x,y), (x^n,y^n), (x^n,y) \in AR^1(t) \ \forall n \in T$.

$(y^n,x), (y,x), (y^n,x^n), (y,x^n) \in AR^2(t) \ \forall n \in T$.

$(x^n,x), (y,y^n) \in AR^3(t) \ \forall n \in T$ and $(x,y) \in SR^3(t)$.

$(x,x^n), (x,y), (y^m,x^n), (y^m,y) \in AR^4(t)$ if $m \leq t$ and $n \leq t$.

$(x^n,y^m) \in AR^4(t)$ if $m > t$ and $n > t$.

$(x^n,x),(y,x),(x^n,y^m),(y,y^m) \in AR^5(t)$ if $m \le t$ and $n \le t$.
$(y^m,x^n) \in AR^5(t)$ if $m > t$ and $n > t$.

This definition is quite contrived; its value arises from the fact that $P(X)^T$ and $W(Y)$ are both elementary because the oligarchy theorem (8.36) goes through when T is infinite and the domain is elementary. (As above, $Y = \{a \in \mathbf{E}^2_{++} \mid \log a_1 + \log a_2 = 0\}^T$ endowed with the product topology.) As with the other domain conditions in Chapters 8 and 9, this definition makes demands on the topology as well as the domain. Although our terminology is a little sloppy, it has the virtue of being succinct. (In particular, the topology cannot be discrete: there exist at least two convergent sequences that are not eventually constant.)

9. Lemma. Let T be any set. If D is standard, $f: D \to Q(X)$ satisfies strict non-imposition and Arrow's independence axiom, and (I,J) is decisive for $f \mid \{x,y\}$ and $\{x,y\}$ is free, then (I,J) is decisive for f.

Proof. Let $\{a,b\}$ be any free pair. By 15d of 18.15 there exist $w,z, \in X$ such that $B = \{x,y,w\}$, $C = \{y,w,z\}$, $E = \{w,z,a\}$, and $F = \{z,a,b\}$ are free triples. Then (I,J) is decisive for $f \mid B$ by Lemma 4.4. Then (I,J) is decisive for $f \mid C, f \mid E$, and $f \mid F$ by virtue of the same lemma. Therefore, (I,J) is decisive over all free pairs. If $(x,y) \in \cap_{i \in I} Ap(i) \cap_{j \in J} -Ap(j)$ choose $z \in X$ such that $\{x,z\}$ and $\{y,z\}$ are free (15c). Choose $p' \in D$ so that $p'(t) = p(t)$ for all $t \notin I \cup J$ and (x,z) and (z,y) belong to $\cap_{i \in I} Ap'(i)$ $\cap_{j \in J} -Ap'(j)$. (Use 15a, 15b, and 15c.) Then $(x,z),(z,y) \in Af(p')$ so $(x,y) \in Af(p)$ by transitivity of $Af(p')$ and the independence axiom. ∎

Three more lemmas are required at this point. We need to prove that a minimal pair of decisive coalitions exists. This is easy if T is finite (Theorem 8.29). More work is required in the case of a denumerable society. Given a social welfare function f let I^* be the intersection of all subsets I of T such that (I,J) is decisive for some $J \subset T$. Similarly, let J^* be the intersection of all subsets J of T such that (I,J) is decisive for some $I \subset T$.

10. Lemma. Let T be the set of positive integers. If D is elementary and $f: D \to Q(X)$ satisfies strict non-imposition and Arrow's independence axiom then there exists a decisive pair of coalitions.

Proof. Let $Z = \{x,y\}$ be the free pair and let $\{x^n\}$, $\{y^n\}$ be the sequences associated with x and y that are guaranteed by the definition of an elementary domain. There is some $p \in D$ such that $(x,y) \in Af(p)$ because f satisfies strict non-imposition. Let $I = \{t \in T \mid (x,y) \in Ap(t)\}$ and $J = \{t \in T \mid (y,x) \in Ap(t)\}$. For each $t \in T$ let $R^1(t)$, $R^2(t)$, and $R^3(t)$ have the properties specified by Definition 8. Define $p' \in D$ by setting $p'(t) = R^1(t)$ if $t \in I$, $p'(t) = R^2(t)$ if $t \in J$, and $p'(t) = R^3(t)$ if $t \notin I \cup J$. We have $(x,y) \in Af(p')$ by Arrow's independence axiom. By continuity of $f(p')$ there is some t such that $(x^t,y^t) \in Af(p')$. Then $(T - J, J)$ is decisive for (x^t,y^t) by the independence axiom. Then $(T - J, J)$ is decisive for f by 8b and Lemma 9. ∎

11. Lemma. Let T be the set of positive integers. If D is elementary and $f: D \rightarrow Q(X)$ satisfies strict non-imposition and Arrow's independence axiom then either I^* or J^* is not empty.

Proof. There is at least one decisive pair (I,J) by Lemma 10. If $I^* = \varnothing = J^*$ then for each $t \in T$ there exist decisive pairs (I',J') and (I'',J'') such that $t \notin I'$ and $t \notin J''$. Let Z be any free triple. Then $(I',J') = (I \cap I' \cap I'', J \cap J' \cap J'')$ is decisive for $f \mid Z$ by Lemma 4.5 and hence for f by Lemma 9. And $t \notin I' \cup J'$. Set $H^t = I^1 \cap I^2 \ldots \cap I^t$ and $K^t = J^1 \cap J^2 \cap \ldots \cap J^t$. Using Lemmas 9 and 4.5 we can show that (H^t,K^t) is decisive for f. Now bring on stage the free pair $\{x,y\}$ and $\{x^t\}$ and $\{y^t\}$ of Definition 8 and define $p \in D$ by setting $p(t) = R^4(t)$ if $t \in I$, $p(t) = R^5(t)$ if $t \in J$, with $p(t)$ arbitrary if $t \notin I \cup J$. For any $t \in T$ we have $(y^t,x^t) \in Ap(n)$ for all $n \in H^{t-1}$ and $(x^t,y^t) \in Ap(n)$ for all $n \in K^{t-1}$. Therefore, $(y^t,x^t) \in Af(p)$ for all $t \in T$ by decisiveness of (H^{t-1},K^{t-1}). But $(x,y) \in Af(p)$ by decisiveness of (I,J), contradicting continuity of $f(p)$ because $\{x^t\}$ converges to x and $\{y^t\}$ converges to y. Therefore, $I^* \cup J^*$ must be non-empty. ∎

12. Lemma. Let T be the set of positive integers. If D is elementary and $f: D \rightarrow Q(X)$ satisfies strict non-imposition and Arrow's independence axiom then both I^* and J^* are finite and (I^*,J^*) is decisive.

Proof. We begin by showing that there exists a finite decisive pair. This will imply that I^* and J^* are finite. There exists a decisive pair (I,J) by Lemma 10. Define $p \in D$ by setting $p(t) = R^4(t)$ if $t \in I$, $p(t) = R^5(t)$ if $t \in J$, and $p(t) = R^3(t)$ if $t \notin I \cup J$. Then

(x,y) belongs to $Af(p)$ because (I,J) is decisive. By continuity of $f(p)$ there is some $n \in T$ such that $(x',y') \in Af(p)$ if $t \geq n$. Set $N = \{1,2,\ldots,n-1\}$, $I' = \{t \in T \mid (x^n,y^n) \in Ap(t)\}$ and $J' = \{t \in T \mid (y^n,x^n) \in Ap(t)\}$. The pair (I',J') is decisive for (x^n,y^n) by the independence axiom and the fact that $I' \cup J' = T$. Then (I',J') is decisive by 8b and Lemma 9. Then $(I'',J'') = (I \cap I', J \cap J')$ is decisive for any free triple by Lemma 4.5 and hence for f as well by Lemma 9. (Because $\{x,y\}$ is free there exists a free triple by 15d.) But I'' and J'' are both contained in N because $t < n$ follows from either $t \in I$ and $(x^n,y^n) \in Ap(t)$ or $t \in J$ and $(y^n,x^n) \in Ap(t)$.

If (H,K) is any decisive pair then $(H \cap I'', K \cap J'')$ is decisive for any free triple by Lemma 4.5, and hence for f as well by Lemma 9. Therefore, $(H, J'' - H)$ is decisive because H contains $H \cap I''$ and $J'' - H$ contains $K \cap J''$. Therefore, $I^* = \cap \{H \subset I'' \mid (H, J'' - H)$ is decisive.$\}$ Because intersection is taken over a finite set the pair (I^*, J'') is decisive by Lemmas 4.5 and 9. I^* is finite as a subset of a finite set. Similarly, (I'', J^*) is decisive. Therefore, $(I^*, J^*) = (I^* \cap I'', J'' \cap J^*)$ is decisive by Lemmas 4.5 and 9. ∎

13. Theorem. If T is countable, D is elementary, and $f \colon D \to Q(X)$ satisfies strict non-imposition and Arrow's independence axiom then f is oligarchical and the oligarchy is finite.

Proof. By Lemma 12, (I^*, J^*) is decisive and $I^* \cup J^*$ is finite. If Z is any free triple then (I^*, J^*) is a minimal decisive pair for $f \mid Z$. Then each $i \in I^*$ (resp. $j \in J^*$) has veto power (resp. inverse veto power) for $f \mid Y$ for each free pair Y by Lemma 4.8. Suppose $t \in I^*$ and $(x,y) \in Ap(t) \cap -Af(p)$. Choose $z \in X$ so that $\{x,z\}$ and $\{z,y\}$ are free. Choose $p' \in D$ so that $(x,y),(z,y) \in Ap'(t)$, $(x,z) \in \cap_{i \in I^*} Ap'(i) \cap_{j \in J^*} -Ap'(j)$, and $p'(i) \cap \{x,y\}^2 = p(i) \cap \{x,y\}^2 \forall i \in T$. We can set $p'(i) = p(i) \forall i \notin I^* \cup J^*$. If $i \in I^* \cup J^*$ but $x \notin F_i(y)$ the desired $p'(i)$ exists because $D_i(\{x,y\})$ is then a singleton by 15a of 18.15. If $x \in F_i(y)$ then $D_i(\{x,y,z\}) = P(\{x,y,z\})$ by 15b. We have $(y,x) \in Af(p')$ by Arrow's independence axiom and $(x,z) \in Af(p')$ because (I^*, J^*) is decisive. Therefore $(y,z) \in Af(p')$ by transitivity of $Af(p')$. This contradicts the fact that t has a veto for $f \mid \{y,z\}$. Therefore, every member of I^* has a veto for f. Similarly, each member of J^* has inverse veto power for f. ∎

14. Lemma. If T is countable and X is a normal and connected Hausdorff space satisfying the first axiom of countability then $P(X)^T$ is elementary.

Proof. As in the proof of Theorem 3 we can assume the existence of sequences $\{x'\}$ in $X - \{x\}$ and $\{y'\}$ in $X - \{y\}$ converging to x and y respectively, and because the space is Hausdorff we can further assume that the points $x, x^1, \ldots, x', \ldots, y, y^1, \ldots, y', \ldots$ are distinct. Assume that T is the set of positive integers.

First, we show that we can find $R^1(t)$ and $R^2(t)$ in $P(X)$ with the desired properties. By Urysohn's theorem (2.9) there exists a continuous real-valued function α on X such that $\alpha(x) = \alpha(x^n) = 1 = 1 + \alpha(y) = 1 + \alpha(y^n)$ for all $n \in T$. Set $(x',y') \in R^1(t)$ if and only if $\alpha(x') \geq \alpha(y')$ and $(x',y') \in R^2(t)$ if and only if $\alpha(x') \leq \alpha(y')$.

To define $R^3(t)$ we use a simple corollary of Theorem 2.9. If F_1, F_2, and F_3 are mutually disjoint closed subsets of X there exist continuous functions $\alpha\colon X \to [0,1]$ and $\beta\colon X \to [0,1]$ such that $\alpha(F_1) = \{1\}$, $\alpha(F_2 \cup F_3) = \{0\}$, $\beta(F_1 \cup F_2) = \{1\}$, and $\beta(F_3) = \{0\}$. Therefore, $\gamma = \alpha + \beta$ is a continuous real-valued function such that $\gamma(F_1) - \gamma(F_2) = 1 = \gamma(F_2) - \gamma(F_3)$. Therefore, for each $n \in T$ there is a continuous function $\gamma_n\colon X \to [-2^{-n}, 2^{-n}]$ such that $\gamma_n(x^n) = 2^{-n}$, $\gamma_n(y^n) = -2^{-n}$, and $\gamma_n(x') = \gamma_n(x) = \gamma_n(y) = \gamma_n(y') = 0$ if $t \neq n$. The function $\gamma = \Sigma_{n \in T} \gamma_n$ is continuous by Proposition 2.12. Set $(x',y') \in R^3(t)$ if and only if $\gamma(x') \geq \gamma(y')$.

For each $t \in T$ there exists a continuous function $\delta\colon X \to [0,1]$ such that

$$\delta(x) = \delta(y^n) = 1 = 1 + \delta(x^n) = 1 + \delta(y) \text{ if } n \leq t.$$
$$\delta(x^n) = 1 = 1 + \delta(y^n) \text{ if } n > t.$$

(Use Theorem 2.9 again.) Set $(x',y') \in R^4(t)$ if and only if $\delta(x') \geq \delta(y')$ and $(x',y') \in R^5(t)$ if and only if $\delta(x') \leq \delta(y')$. The domain $P(X)^T$ is standard by Lemma 8.12. ∎

15. Theorem. Assume that T is countable and X is a normal and connected Hausdorff space satisfying the first axiom of countability. Then $f\colon P(X)^T \to Q(X)$ satisfies Arrow's independence axiom and minimal responsiveness if and only f is completely oligarchical and the oligarchy is finite.

Proof. (i) Suppose that $f\colon P(X)^T \to Q(X)$ is completely oligarchical, (I,J) is the oligarchy underlying f, and I and J are both finite. A completely oligarchical social welfare function satisfies Arrow's independence axiom without restriction and it satisfies minimal responsiveness because $P(X)^T$ has the free-triple property. If $(x,y) \in Af(p)$ then $(x,y) \in Ap(i) \cap -Ap(j)$ for all $i \in I$ and $j \in J$. Then for each $t \in I \cup J$ there exist neighbourhoods $N_t(x)$ and $N_t(y)$ of x and y respectively such that $(x',y') \in Ap(i) \cap -Ap(j)$ for all $x' \in N_i(x) \cap N_j(x)$, all $y' \in N_i(y) \cap N_j(y)$, and all $i \in I$, $j \in J$. Because I and J are finite the sets $N(x) = \cap_{t \in I \cup J} N_t(x)$ and $N(y) = \cap_{t \in I \cup J} N_t(y)$ are open. We have $(x',y') \in Af(p)$ for all $x' \in N(x)$, $y' \in N(y)$ because (I,J) is decisive. Then $f(p)$ is continuous. It is obviously quasitransitive.

(ii) Suppose that $f: P(X)^T \to QP(X)$ satisfies Arrow's independence axiom and minimal responsiveness. Then f satisfies strict non-imposition by part (ii) of the proof of Theorem 8.35. (Minimal responsiveness implies strict non-imposition for any T.) Therefore f is oligarchical and the oligarchy is finite by Theorem 13 and Lemma 14. It is completely oligarchical by Lemma 8.13 and Proposition 8.10. ∎

16. Example. T is the set of positive integers and $X = \mathbf{E}_+$, the nonnegative part of the real line. $D = P(X)^T$. Set $(x,y) \in Af(p)$ if and only if $(x,y) \in Ap(t)$ for all $t \in T$. (This the Pareto aggregation rule.)

Define the profile $p \in D$ by means of a continuous real-valued function u_t on $\mathbf{E}_+$ for each $t \in T$.

$u_t(x) = x$ if $x \le 1$.
$u_t(x) = 2 - x + (x - 1)(1 - 1/t)$ if $1 \le x \le 2$.
$u_t(x) = 1 - 1/t + t(x - 2)$ if $x \ge 2$.

We have $(1,2) \in Af(p)$ because $u_t(1) > u_t(2)$ for all $t \in T$. For every $\varepsilon > 0$ we have $u_t(x) > u_t(1)$ for all $x > 2 + \varepsilon/2$ and $t > (2/\varepsilon)^{1/2}$. Therefore $(x,1) \in f(p)$ for all $x > 2$, and thus $f(p)$ is not continuous.

The results of this chapter are extended to allocation spaces and economic environments in Chapter 12. The overlapping generations model is featured.

Part IV
Economic Environments

Introduction

Chapters 10 and 11 establish the impossibility theorems of Chapter 8 for the realm of classical welfare economics. Private goods and the restricted domain of selfish and monotonic individual preferences are treated in Chapter 10. Chapter 11 considers a model with public goods, and then a full model with public and private goods. Preferences are appropriately restricted, of course. In short, we not only extend the classical impossibility theorems on dictatorship, inverse dictatorship, and oligarchy to families of economic environments, we also use the ambient algebraic and topological structure to prove very strong versions of those theorems.

Kalai *et al.* (1979) first proved the Arrow impossibility theorem for economic environments. They employed a model with public goods only. This paper not only inspired new theorems, its method of proof has been widely used as a model for extensions and refinements, although the next pathbreaking paper on social choice in economic environments, Border (1983), is an exception. Border proved Wilson's theorem for the realm of pure private goods. The proof is very elegant, but it does not appear to suggest similar techniques for other models or theorems. Bordes and Le Breton (1989) showed how the technique of Kalai *et al.* could be strengthened and adapted to prove the classical dictatorship and oligarchy theorems for generalized families of economic environments. Campbell (1989*a,b*) shows that the boundary of the allocation space is in the thrall of the Arrow and Wilson theorems when effectivity and continuity of social preference, respectively, are assumed.

Redekop (1991) proves that Arrow's theorem is in a sense generic for economic models. Any domain of individual preference profiles defined over allocation spaces will be very contrived (nowhere dense in the Kanai topology) if it admits a non-dictatorial social welfare function satisfying the Pareto criterion, Arrow's independence axiom, and full transitivity of social preference. Redekop proves a corresponding result for quasitransitive social preference. Nagahisa

(1991) extends some central results on *acyclic* social choice to economic environments.

Donaldson and Roemer (1987) and Donaldson and Weymark (1988) consider economic environments and they take social choice theory in new directions. The former analyses the implications for collective choice when a natural restriction is placed on the way that the social choice rule responds to a change in the number of commodities. Donaldson and Weymark treat the case of realistically defined feasible sets — sets that are convex in particular.

10 Private Goods

THIS chapter extends the theorems of Chapter 8 on the impossibility of efficiency–equity trade-offs to the realm of classical welfare economics with private goods and monotonic and self-regarding (or selfish) individual preferences. In this setting dictatorship results in everyone but the dictator receiving zero units of every good. To be specific, suppose that t is a dictator and the feasible set Z is closed under redistribution. In other words, x belongs to Z whenever y belongs to Z and $\Sigma_{t \in T} x(t) = \Sigma_{t \in T} y(t)$. If $y \in Z$ and $y(i) \neq 0$ for $i \neq t$ we have $(x,y) \in Ap(t)$ for $x(t) = y(t) + y(i)$ because $p(t)$ is monotonic and independent of $x(j)$ for $j \neq t$. Therefore, everyone but the dictator is assigned the zero commodity vector by any allocation in Z that is maximal with respect to $f(p)$. Obviously, dictatorship is far more objectionable in this context than when a committee must decide on a date for the next meeting, say. If t is an *inverse* dictator then $f(p) = -p(t)$ and x does not maximize $f(p)$ in Z unless $x(t) = 0$. What is worse, the zero allocation, which sets $x(i) = 0$ for all $i \in T$, maximizes $f(p)$ in Z whenever it belongs to Z. Both dictatorship and inverse dictatorship violate everyone's notion of equity. But if social preference is continuous and transitive then the only other rules that are consistent with Arrow's independence axiom are constant social welfare functions. A constant social welfare function is completely unresponsive to individual preference and hence has no efficiency content. With transitive social preference, a social welfare function satisfying Arrow's independence axiom will either have zero equity content or zero efficiency content. There is no middle ground.

If social preference is merely required to be quasitransitive then the Pareto aggregation rule qualifies and it satisfies the Pareto criterion and Arrow's independent axiom. Although it is consistent with non-zero consumption by every person it is completely silent on matters of equity, or distributional justice. Because it is the only social welfare function that satisfies Arrow's independence axiom and strict non-imposition, and does not either ignore someone's preference scheme or incorporate it into the social ranking in a perverse way, we can also claim that there are no equity–efficiency trade-offs in the case of quasitransitive social preference.

We prove these claims for the finite society case in this chapter, and for the infinite society, overlapping generations model in Chapter 12.

Recall that Ω is the basic allocation space of k private goods. $\Omega = (\mathbf{E}_+^k)^\mathrm{T}$. The classical domain is W, the family of profiles of continuous, self-regarding, monotonic, and convex preorders. $W = \Pi_{t \in \mathrm{T}} W_t$.

1. The classical domain. W_t is the set of continuous preorders $R \in P(\Omega)$ such that for all $w,x,y,z \in \Omega$ the following three conditions hold:

1a $w(t) = x(t)$ and $y(t) = z(t)$ imply wRz if and only if xRy (self-regarding preferences).

1b $x(t) \geq y(t) \gg 0$ and $x(t) \neq y(t)$ implies $(x,y) \in AR$ (strict monotonicity).

1c the set $\{v \in \Omega \mid vRx\}$ is convex.

Constant social welfare functions are completely unresponsive to individual preference. However, there exists a constant social welfare function *on the domain* W that generates Pareto optimal outcomes. Simply choose some $t \in T$ and set $(x,y) \in f(p)$ if and only if $x(t) \geq y(t)$. If $\omega \in \mathbf{E}_{++}^k$ and $Z = \{x \in \Omega \mid \Sigma_{t \in \mathrm{T}} x(t) \leqslant \omega\}$ we have $\{x \in Z \mid (y,x) \in Af(p) \rightarrow y \notin Z\} = \{x \in Z \mid (y,x) \in Ap(t) \rightarrow y \notin Z\}$ for all $p \in W$ because $p(t)$ is self-regarding and monotonic.

When public goods are added to the model (Chapter 11) or Z is not closed under redistribution then this trick of embodying efficiency in a constant social welfare function will not work. Therefore, we are justified in saying that a constant social welfare function has no efficiency content if there are two or more private goods. Of course, W is a singleton when there is only one private good ($k = 1$). In that case Arrow's independence axiom is satisfied vacuously and we can easily define a non-dictatorial social welfare satisfying the Pareto criterion. For example, set $(x,y) \in f(p)$ if and only if $\Sigma_{t \in \mathrm{T}} x(t) \geqslant \Sigma_{t \in \mathrm{T}} y(t)$. If $(x,y) \in Ap(t)$ for all $t \in T$ then $x(t) > y(t)$ for all $t \in T$ and thus $(x,y) \in Af(p)$. Because W is a singleton when $k = 1$ any social welfare function is constant and we can extend Theorem 8.26 to the domain W for any k. The oligarchy theorem (8.36) is valid only for $k \geqslant 2$, and we will concentrate on the case $k = 2$ in this chapter.

We begin by pointing out that Arrow's theorem is false for the set of allocations of private goods and the domain W if we do not require continuity of preference in any form. This was demonstrated in Blau (1957) and we now reproduce Blau's example.

2. Blau's example. Set $T = \{1,2\}$ and for $S \subset T$ let $X^S = \{x \in \Omega \,|\, x(t) = 0 \,\forall\, t \in S \text{ and } x(t) \neq 0 \,\forall\, t \notin S\}$. Assume the discrete topology for Ω and define $f: W \rightarrow P(\Omega)$ as follows:

2a Set $(x,y) \in Af(p)$ whenever $x \in X^I$, $y \in X^J$, and I is a proper subset of J.

2b Let $f(p)$ coincide with $p(1)$ on $X^\emptyset$ for all $p \in W$.

2c Let $f(p)$ coincide with $p(1)$ on $X^{\{2\}}$ for all $p \in W$.

2d Let $f(p)$ coincide with $p(2)$ on $X^{\{1\}}$ for all $p \in W$.

This social welfare function is non-dictatorial. Although 1 dictates $f\,|\,X^\emptyset$ we have $(x,y) \in Af(p)$ if $x(1) = (1,1, \ldots,1) = x(2)$, $y(1) = (2,2, \ldots,2)$, and $y(2) = (0,0, \ldots,0)$. It obviously satisfies Arrow's independence of irrelevant alternatives condition. It satisfies the Pareto criterion even though $f\,|\,\{x,y\}$ is constant for many pairs $\{x,y\}$. If $x \in X^I$, $y \in X^J$, and I is a proper subset of J then $y(t) = 0 \neq x(t)$ for $t \in J - I$ which means that $(y,x) \in \cap_{t \in T} Ap\,(t)$ does not hold for any $p \in W$ and there is no violation of the Pareto criterion.

It is no accident that $f\,|\,X^\emptyset$ is dictatorial. Let Ω_0 denote the set of allocations x in Ω such that $x(t) \gg 0$ for all $t \in T$. That is, $\Omega_0 = \{x \in \Omega \,|\, x_c(t) > 0 \,\forall\, t \in T \text{ and } c = 1,2, \ldots,k\}$. Assuming the discrete topology Border (1983) and Bordes and Le Breton (1989) prove that $f: W(\Omega_0) \rightarrow P(\Omega_0)$ is dictatorial if it satisfies Arrow's independence axiom and the Pareto criterion, even if the discrete topology is employed. In other words, if $f(p)$ is a preorder for all $p \in W(\Omega_0)$ and f satisfies Arrow's independence axiom and the Pareto criterion then f is dictatorial. (These two papers also prove Wilson's theorem for the outcome set Ω_0 and domain $W(\Omega_0)$.) Blau showed how the boundary could be added and used to break the grip of dictatorship without violating the Pareto criterion. Border and Bordes–Le Breton prove that any such social welfare function will be dictatorial on Ω_0, the set of strictly positive allocations. Because the allocations of interest in the real world are on the boundary $\Omega - \Omega_0$ we need to investigate further. One reason why Blau's example is unsatisfactory is that the choice set is almost always empty. Let $Z \subset \Omega$ be any set of two or more allocations. Suppose that y belongs to Z whenever x does and $y(1) = \frac{1}{2}x(1) + \frac{1}{2}x(2) = y(2)$ or $y(1) = x(1) + \frac{1}{2}x(2)$ and $y(2) = \frac{1}{2}x(2)$. Now, choose $x \in Z - \{0\}$ and set $y(t) = \frac{1}{2}x(1) + \frac{1}{2}x(2)$ for $t = 1,2$. Set $R = f(p)$. We have $y \in X^\varnothing$ so $\max R(Z) \subset X^\varnothing$. (Recall Defnition 1.11.) If $x \in Z \cap X^\varnothing$ then $z \in X^\varnothing$ for $z(1) = x(1) + \frac{1}{2}x(2)$ and $z(2) = \frac{1}{2}x(2)$. Then $(z,x) \in Af(p)$ for all $p \in W$. Therefore, $x \notin \max R(Z)$ and we have $\max R(Z) = \varnothing$ for all $p \in W$. In words, there is no socially best alternative if the feasible set allows for redistribution. Our first theorem shows that this is unavoidable if f is a non-dictatorial social welfare function on W satisfying the Pareto criterion, Arrow's independence axiom, and transitivity of social preference. First we need to formalize the idea that $\max R(Z)$ is not empty for realistic feasible sets.

A subset Z of Ω is *compact* if it is closed and bounded. Therefore, a set comprised of a sequence and its limit point is compact. The binary relation R on Ω is said to be *effective* if $\max R(Z) \neq \varnothing$ for every non-empty and compact subset Z of Ω. The next theorem is independent of the other results in this book. The reader who is only concerned with efficiency–equity trade-offs is advised to skip to Lemma 4.

3. Theorem. Let f be a social welfare function for the outcome set Ω and domain W. Suppose that $f(p)$ is an effective preorder for each $p \in W$. Then f is dictatorial if it satisfies the Pareto criterion and Arrow's independence axiom.

Proof. $f \mid \Omega_0$ is dictatorial, even if effectiveness is not imposed. This is proved in Border (1983) and in Bordes and Le Breton (1989). Let t be the dictator for $f \mid \Omega_0$. Suppose that $x,y \in \Omega$ and (x,y) belongs to $Ap(t)$. We prove that (x,y) belongs to $Af(p)$. First, suppose $y \in \Omega_0$. By Lemma 2.19 there is some $z \in \Omega$ such that $(x,z),(z,y) \in Ap(t)$. By continuity of $p(t)$ there is some neighbourhood $N(z)$ of z such that $(x,z'),(z',y) \in Ap(t)$ for all $z' \in N(z)$. Choose $z' \in N(z) \cap \Omega_0$. Then $(z',y) \in Af(p)$. If $(y,x) \in f(p)$ we have $(z',x) \in Af(p)$ by transitivity (1.27). Now, define the infinite sequence $\{z^0, z^1, z^2, \dots\}$ in Ω_0. Set $z^0 = z'$. To define z^n inductively assume that $(x,z^{n-1}) \in Ap(t)$. There is some neighbourhood $N(x)$ of x such that $(v,z^{n-1}) \in Ap(t)$ for all

$v \in N(x)$. Choose z^n so that it belongs to $N(x) \cap \{z \in \Omega_0 \mid$ the distance between x and z does not exceed $n^{-1}\}$. Then $\{z^n\}$ converges to x. Therefore, the set $Z = \{x, z^0, z^1, z^2, \ldots\}$ is compact. Set $R = f(p)$. We have $x \notin \max R(Z)$ because $(z^0, x) \in Af(p)$. And $z^n \notin \max R(Z)$ for any n because $(z^{n+1}, z^n) \in Ap(t)$, z^{n+1} and z^n belong to Ω_0, and t dictates $f \mid \Omega_0$. Therefore, $\max R(Z) = \varnothing$, a contradiction. Therefore, (i) we must have $(x, y) \in Af(p)$ whenever $(x, y) \in Ap(t)$ and $y \in \Omega_0$.

Suppose $(x, y) \in Ap(t)$ and $x, y \in \Omega$. Choose $z \in \Omega_0$ such that $(x, z), (z, y) \in Ap(t)$. Now define the sequence $\{z^0, z^1, z^2, \ldots\}$ in Ω_0. Set $z^0 = z$ and with z^{n-1} defined set $z^n(i) = y(i) + (n^{-1}, n^{-1})$ for $i \neq t$ and choose integer $m \geq n$ large enough so that $z^n(t) = y(t) + (m^{-1}, m^{-1})$ satisfies $(z, z^n) \in Ap(t)$. This is possible because (z, y) belongs to $Ap(t)$ and $p(t)$ is continuous. Set $Z = \{y, z^0, z^1, z^2, \ldots\}$, which is compact because $\{z^n\}$ converges to y. But $z^n \notin \max R(Z)$ for $n \geq 1$ because z^0 and z^n both belong to Ω_0 and $(z^0, z^n) \in Ap(t)$. And $y \notin \max R(Z)$ by the Pareto criterion and monotonicity of individual preference: we have $(z^1, y) \in Af(p)$. Therefore, $\max R(Z) = \{z^0\} = \{z\}$ by effectiveness. Therefore, $(z, y) \in Af(p)$. We have $(x, z) \in Af(p)$ by (i) and thus $(x, y) \in Af(p)$ by transitivity. ∎

The hypothesis of Theorem 3 does not include the assumption that social preferences are continuous. In other words, the discrete topology may be assumed in evaluating continuity of social preference, although the proof assumes that *individual* preferences are continuous in the Euclidean topology and compactness is defined in terms of that topology. Effectiveness and the Pareto criterion substitute for continuity of social preference in providing leverage for the proof.

Now we return to the assumption that social preference is continuous in a non-trivial topology; we maintain that assumption for the rest of the book. Initially we focus attention on $f \mid \Omega_0$. For transitive social preferences the impossibility theorem extends immediately to f itself by virtue of the following unique-extension lemma.

4. Lemma. Let Y be a connected subspace of the topological space X. If R and R' are continuous preorders on the closure of Y, and $R \cap Y^2 = R' \cap Y^2$ then $R = R'$.

Proof. (i) $(x, y) \in AR \cap -AR'$ cannot hold. Suppose

$(x,y) \in AR \cap -AR'$. By continuity of R and R' there exist neighbourhoods $N(x)$ and $N(y)$ of x and y respectively such that $N(x) \times N(y) \subset AR \cap -AR'$. Because x and y belong to the closure of Y there exist $x' \in N(x) \cap Y$ and $y' \in N(y) \cap Y$. Then $(x',y') \in AR \cap -AR'$, contradicting $R \cap Y^2 = R' \cap Y^2$.

(ii) $x \in Y$ and $(x,y) \in AR$ imply $(x,y) \in AR'$. Suppose $(x,y) \in AR \cap - R'$ and $x \in Y$. By connectedness of the closure of Y (Lemma 2.17) there exists some z in the closure of Y such that $(x,z),(z,y) \in AR$ (2.19). By continuity of R there is a neighbourhood $N(z)$ of z such that $(x,z'),(z',y) \in AR \; \forall z' \in N(z)$. Choose any $z' \in N(z) \cap Y$. Then $(x,z') \in AR'$ by hypothesis. Because $(y,x) \in R'$ we have $(y,z') \in AR'$ by 1.27. This contradicts (i). Therefore, $x \in Y$ and $(x,y) \in AR$ imply $(x,y) \in AR'$.

Suppose $(x,y) \in AR \cap -R'$. Choose $z \in Y$ such that $(x,z),(z,y) \in AR$. We have $(z,y) \in AR'$ by (ii) and hence $(z,x) \in AR'$ by 1.27, contradicting (i). ■

We cannot prove a unique-extension lemma for continuous quasiorders as the next example demonstrates. (It also shows that continuity is essential to Lemma 4.)

5. Example. Define R, R', and R'' on $\mathbf{E}_+^2$ by setting

xRy if and only if $x_1 + x_2 \geq y_1 + y_2$.

$xR'y$ if and only if $[x_1 + x_2 \geq y_1 + y_2$ and $x_1 x_2 \neq 0]$ or $[y_1 y_2 = 0]$.

$xR''y$ if and only if $[x_1 + x_2 \geq y_1 + y_2$ and $x_1 x_2 y_1 y_2 \neq 0]$ or $[x_1 x_2 y_1 y_2 = 0]$.

Obviously R, R', and R'' coincide on $\mathbf{E}_{++}^2$ and the restriction of each to $\mathbf{E}_{++}^2$ is a continuous preorder. But $R \neq R' \neq R'' \neq R$. R' is not continuous because $(1,1)$ is strictly preferred to $(0,5)$ according to R' but every neighbourhood of $(0,5)$ contains points strictly preferred to $(1,1)$. R' is a preorder on $\mathbf{E}_+^2$ that is not continuous. R'' is continuous on $\mathbf{E}_+^2$ because $(x,y) \in AR''$ implies $x,y \in \mathbf{E}_{++}^2$ and therefore there are neighbourhoods $N(x)$ and $N(y)$ such that $a_1 + a_2 > b_1 + b_2$ *and* $a,b \in \mathbf{E}_{++}^2$ for all $a \in N(x)$ and $b \in N(y)$. But R'' is not a preorder: $(1,1)$ is indifferent to $(1,0)$ which is indifferent to $(2,2)$ but $(2,2)$ is strictly preferred to $(1,1)$.

Assume $k = 2$ for the rest of the chapter. Now we construct connected subspaces Y of Ω such that $W(Y)$ is standard (8.15), i.e. $W(Y)$ 'almost' has the free-triple property. These subspaces are products of *strictly convex downward-sloping curves*, called SCDS curves from now on.

6. SCDS curve. A strictly convex downward-sloping curve is a subset C of $\mathbf{E}^2_{++}$ with the following properties: There is a continuous function $h: (0,1) \to \mathbf{E}^2_{++}$ such that $C = h((0,1))$, and conditions 6a–6c are met, with 6a and 6b holding for all distinct $x, y, z \in C$.

6a $\quad (x_1 - y_1)(x_2 - y_2) < 0$ if $x \ne y$.

6b $\quad (x_2 - y_2)(z_1 - y_1) > (y_2 - z_2)(y_1 - x_1)$ if $x_1 < y_1 < z_1$.

6c $\quad$ For any integer $n > 0$ there exist $v, w \in C$ such that $v_1 > n$ and $w_2 > n$.

An SCDS curve is a connected subspace of $\mathbf{E}^2$ by Lemma 2.14 and thus the product of SCDS curves is connected by Lemma 2.18. An SCDS curve is downward sloping because $y_1 > x_1$ implies $y_2 < x_2$. It is strictly convex because $x_1 < y_1 < z_1$ implies $-(y_2 - x_2)/(y_1 - x_1) > -(z_2 - y_2)/(z_1 - y_1)$. Condition 6c is not really necessary but it simplifies the argument at a couple of points. It ensures that the curve is unbounded.

The next lemma does most of the work of proving that $W(Y)$ is standard if Y is a product of SCDS curves. The reader may be content to experiment with a diagram or two instead of working through Lemma 7. For this lemma only we think of the members of W_t as binary relations on $\mathbf{E}^2_+$; this is justifiable because the members of W_t are independent of every component of $x \in [\mathbf{E}^2_{++}]^{\mathsf{T}}$ other than $x(t)$.

7. Lemma. If $C \subset \mathbf{E}^2_{++}$ is an SCDS curve then for any $t \in T$ and any three-element subset Z of C and any preorder R on Z there is some $R' \in W_t$ such that $R' \cap Z^2 = R$.

Proof. Let Z be any three-element subset of C. There are thirteen preorders on Z so we treat thirteen cases. In each case a continuous preorder R' on $\mathbf{E}^2_+$ is generated by specifying a continuous utility function u on $\mathbf{E}^2_+$. It will be obvious that R' agrees with the given preorder R and Z. We will also have to verify 1b and 1c. In some cases u is a simple linear function with positive coefficients in which case 1b and 1c are obvious. For the other cases u is generated by specifying a single indifference curve (or level curve) Y which will be an SCDS curve. A utility function is derived from Y as follows.

For each $x \in E_+^2$ let $\alpha(x)$ denote the real number for which $\alpha(x)x \in Y$ and set $u(x) = -\alpha(x)$. Certainly, u is continuous. To prove strict monotonicity, suppose that $x \geq y \neq x$. Then $\alpha(y)x \geq \alpha(y)y$. Therefore, $\alpha(y)x$ lies above Y because $\alpha(y)y$ is on Y which is strictly downward sloping. Then $\alpha(x) < \alpha(y)$. Now prove convexity. Suppose that $u(x) \geq u(v)$, $u(y) \geq u(v)$, and $z = \lambda x + (1 - \lambda)y$ for $0 \leq \lambda \leq 1$. Without loss of generality suppose that $u(x) \geq u(y)$. Then $\alpha(x) \leq \alpha(y)$. Set $x' = \alpha(x)x/\alpha(y)$. Then $x \geq x'$ and $\alpha(y)x' \in Y$. Set $z' = \lambda x' + (1 - \lambda)y$. We have $z \geq z'$ and $\alpha(y)z' = \lambda\alpha(y)x' + (1 - \lambda)\alpha(y)y$. Then $\alpha(y)z'$ is a convex combination of points in Y and therefore $\alpha(y)z'$ lies above Y unless $\alpha(y)x' = \alpha(y)y$, in which case $\alpha(y)z'$ coincides with those two points. In either case we have $\alpha(z') \leq \alpha(y)$ and thus $u(z') \geq u(y)$. Because $z \geq z'$ we also have $u(z) \geq u(z')$ by strict monotonicity. Therefore $u(z) \geq u(y) \geq u(v)$.

Set $Z = \{x, y, z\}$ with $x_1 < y_1 < z_1$. Rather than specifying each R on Z we will refer to the implied inequalities for u directly.

Case 1. $u(x) = u(y) = u(z)$. Set $Y = C$.

Case 2. $u(x) > u(y) > u(z)$. Define the linear utility function u by setting $u(v) = \epsilon v_1 + v_2$ with $\epsilon > 0$ small enough to induce the desired inequality.

Case 3. $u(z) > u(y) > u(x)$. Set $u(v) = v_1 + \epsilon v_2$ with ϵ sufficiently small.

Case 4. $u(x) > u(z) > u(y)$. If $u(v) = \alpha_1 v_1 + \alpha_2 v_2$ and α_1 and α_2 are the coefficients of a straight line passing through z and above y but below x the inequalities are satisfied.

Case 5. $u(z) > u(x) > u(y)$. This time use the coefficients of a line passing through x and above y but below z.

Case 6. $u(y) > u(z) > u(x)$. Form the SCDS curve Y by perturbing C slightly so that it lies below y and above x but goes through z.

Case 7. $u(y) > u(x) > u(z)$. Form the SCDS curve Y by perturbing C slightly so that it passes through x but lies above z and below y.

Case 8. $u(x) = u(y) > u(z)$. Form the SCDS curve Y by perturbing C slightly so that it still goes through x and y but lies above z.

Case 9. $u(y) = u(z) > u(x)$. This is analogous to Case 8.

Case 10. $u(x) = u(z) > u(y)$. Set $u(v) = \alpha_1 v_1 + \alpha_2 v_2$, where

α_1 and α_2 are the coefficients of the straight line passing through x and z.

Case 11. $u(x) > u(y) = u(z)$. Let $u(v) = \alpha_1 v_1 + \alpha_2 v_2$, where α_1 and α_2 are the coefficients of the straight line passing through y and z.

Case 12. $u(z) > u(x) = u(y)$. Set $u(v) = \alpha_1 v_1 + \alpha_2 v_2$, where α_1 and α_2 are the coefficients of the straight line passing through x and y.

Case 13. $u(y) > u(x) = u(z)$. Form the SCDS curve Y by perturbing C slightly so that it still goes through x and z but lies below y. ∎

8. Lemma. If $Y \subset \Omega_0$ is the product of SCDS curves then $W(Y)$ is standard.

Proof. Assume that $Y = \cap_{t \in T} Y_t$ and each Y_t is an SCDS curve. Let F_t be defined with respect to the domain $W(Y)$. (See the paragraph preceding Definition 8.15.) By Lemma 7, $x \notin F_t(y)$ implies $x(t) = y(t)$ and thus $W_t(\{x,y\})$ contains only the null preorder (15a). If $x \in F_t(y)$, $y \in F_t(z)$, and $z \in F_t(x)$ then $x(t) \neq y(t) \neq z(t) \neq x(t)$ and hence $W_t(\{x,y,z\}) = P(\{x,y,z\})$ by Lemma 7 (15b). For any $x,y \in Y$ choose $z \in Y$ such that $x(i) \neq z(i) \neq y(i) \forall i \in T$. Then $z \in F(x) \cap F(y)$ by Lemma 7 (15c).

If $\{x,y\}$ and $\{a,b\}$ are free subsets of Y and N and M are non-empty open subsets of Y choose $w \in N$ and $z \in M$ so that $\{x(t),y(t),w(t)\}$, $\{y(t),w(t),z(t)\}$, $\{w(t),z(t),a(t)\}$, and $\{z(t),a(t),b(t)\}$ are all three-element sets for each $t \in T$. Then 15d holds by Lemma 7. ∎

The most useful subspaces are members of the Cobb–Douglas family. To define a typical member choose $\alpha \in \mathbf{E}_{++}^T$. That is, choose a positive real number $\alpha(t)$ and any real $\beta(t)$ for each $t \in T$. Set $Y_t = \{x(t) \in \mathbf{E}_{++}^2 \mid \log x_1(t) + \alpha(t) \log x_2(t) = \beta(t)\}$.

Then $Y(\alpha,\beta) \equiv \Pi_{t \in T} Y_t$ is a product of SCDS curves. Recall that $y >_f z$ represents the fact that $(y,z) \in Af(p)$ holds for some $p \in W$. (4.1.) And $f \mid Z$, for arbitrary $Z \subset \Omega$, is defined by setting $f \mid Z(p) = f(q) \cap Z^2$, where p is a profile of continuous individual preorders on Z and $q \in W$ is any profile for which $q(t) \cap Z^2 = p(t) \forall t \in T$ (8.16). We can prove that $f \mid Y(\alpha,\beta)$ is authoritarian for each choice of α and β if $f: W \to P(\Omega)$ satisfies Arrow's independence axiom and is not constant. This suggests that

f itself is authoritarian, and it is easy to prove this final step. We remind the reader that Ω is the Euclidean space of (non-negative) allocations of two private goods and Ω_0 is the subspace of strictly positive allocations.

9. Lemma. If $f: W \to P(\Omega)$ satisfies Arrow's independence axiom and $f|\Omega_0$ is not constant then there exist $\alpha,\beta \in \mathbf{E}^T$ such that $f \mid Y(\alpha,\beta)$ is authoritarian.

Proof. If $f|\Omega_0$ is not constant then there exist $x,y \in \Omega_0$ and $p,p' \in W$ such that $(x,y) \in Af(p)$ and $(y,x) \in f(p')$. Set $S = \{t \in T \mid$ neither $x(t) \geq y(t)$ nor $y(t) \geq x(t)$ holds$\}$. $S \neq \varnothing$ because $f \mid \{x,y\}$ is not constant. For each $t \in S$ choose $\alpha(t) > 0$ so that $\log x_1(t) + \alpha(t)\log x_2(t) = \log y_1(t) + \alpha(t)\log y_2(t)$ and set $\beta(t) = \log x_1(t) + \alpha(t)\log x_2(t)$. For $t \in S$ set $Y_t = \{a \in \mathbf{E}^2_{++} \mid \log a_1 + \alpha(t)\log a_2 = \beta(t)\}$. For $t \in T - S$ set $Y_t = \{\lambda x(t) + (1-\lambda)y(t) \mid 0 \leq \lambda \leq 1\}$. Then $Y = \Pi_{t \in T} Y_t$ is connected by Lemmas 2.14 and 2.18. By Lemma 2.19 there is some $z \in Y$ such that $(x,z),(z,y) \in Af(p)$. By continuity of $f(p)$ we can assume that $y(t) \neq z(t) \neq x(t)$ if $y(t) \neq x(t)$. Choose $p'' \in W$ such that $p''(t) \cap \{z,y\}^2 = p(t) \cap \{z,y\}^2$ and $p''(t) \cap \{y,x\}^2 = p'(t) \cap \{y,x\}^2 \forall t \in T$. If $x(t) \geq y(t)$ then $x(t) \geq z(t) \geq y(t)$ and if $y(t) \geq x(t)$ then $y(t) \geq z(t) \geq x(t)$ holds. In either case $W_t(\{x,y,z\})$ is a singleton and we can set $p''(t) = p(t)$. If $x(t) \geq y(t)$ and $y(t) \geq x(t)$ are both false then $t \in S$ and $x(t) \neq y(t) \neq z(t) \neq x(t)$. Therefore, $W_t(\{x,y,z\}) = P(\{x,y,z\})$ and we can obviously find the required p''. Thus we have $(z,y) \in Af(p'')$ and $(y,x) \in f(p'')$ by Arrow's independence axiom. Therefore, $(z,x) \in Af(p'')$ by Lemma 1.27. Therefore, $x >_f z >_f x$. By continuity of $f(p)$ and $f(p'')$ there exist neighbourhoods $N(x)$ and $N''(x)$ of x such that $(x',z) \in Af(p) \cap -Af(p'')$ for all $x' \in N(x) \cap N''(x)$.

Now define $Y(\alpha,\beta)$ by setting $\alpha(t) = 1$ and $\beta(t) = \log x_1(t) + \log x_2(t)$ for $t \notin S$. We have $x \in Y(\alpha,\beta)$. Choose w in $Y(\alpha,\beta) \cap N(x) \cap N''(x)$ such that for all $t \in T$ we have:

9a $\quad x(t) \neq w(t) \neq z(t)$.

9b $\quad x(t) \geq z(t) \neq x(t)$ implies $w(t) \geq z(t)$.

9c $\quad z(t) \geq x(t) \neq z(t)$ implies $z(t) \geq w(t)$.

Choose $p^1 \in W$ such that $p^1 \cap \{x,z\}^2 = p(t) \cap \{x,z\}^2$ and $p^1(t) \cap \{z,w\}^2 = p''(t) \cap \{z,w\}^2$. If $x(t) = z(t)$ we can set $p^1(t) = p''(t)$. If $x(t) \neq z(t)$ and $t \notin S$ then $p^1(t) = p(t)$

satisfies the requirement by 9b and 9c. Otherwise, $x(t) \neq z(t)$ and $t \in S$ in which case $p^1(t)$ exists by 9a and Lemma 7. Then (x,z) and (z,w) belong to $Af(p^1)$ by Arrow's independence axiom. Then $(x,w) \in Af(p^1)$ by transitivity; we have $x >_f w$. Similarly, there exists $p^2 \in W$ such that $p^2(t) \cap \{z,w\}^2 = p(t) \cap \{z,w\}^2$ and $p^2(t) \cap \{z,x\}^2 = p''(t) \cap \{z,x\}^2 \forall t \in T$. Then $(w,z),(z,x) \in Af(p^2)$ by independence and hence $w >_f x$ by transitivity. We have $x,w \in Y(\alpha,\beta)$ and $x >_f w >_f x$. Therefore $f \mid Y(\alpha,\beta)$ is not constant. $Y(\alpha,\beta)$ is a connected T_1-space. And $W(Y(\alpha,\beta))$ is standard by Lemma 8. Obviously $f \mid Y(\alpha,\beta)$ satisfies Arrow's independence axiom. Therefore, $f \mid Y(\alpha,\beta)$ is authoritarian by Theorem 8.26. ■

Now we prove that a dictator (resp. inverse dictator) for $f \mid Y(\alpha,\beta)$ must also be a dictator (resp. inverse dictator) for $f \mid Y(\alpha',\beta')$ if $Y(\alpha,\beta)$ intersects $Y(\alpha',\beta')$.

10. Lemma. Suppose that $f:W \to P(\Omega)$ satisfies Arrow's independence axiom. If t is a dictator for $f \mid Y(\alpha,\beta)$ and $Y(\alpha',\beta') \cap Y(\alpha,\beta) \neq \emptyset$ then t is a dictator for $f \mid Y(\alpha',\beta')$.

Proof. Suppose $x \in Y(\alpha',\beta') \cap Y(\alpha,\beta)$. For each $t \in T$ define the SCDS curve $Y_t \subset \mathbf{E}_{++}^2$ as follows:

To the left of $x(t)$ the curve Y_t coincides with whichever of the curves $\{a \in \mathbf{E}_{++}^2 \mid \log a_1 + \alpha(t)\log a_2 = \beta(t)$ and $a_1 \leq x_1(t)\}$ and $\{a \in \mathbf{E}_{++}^2 \mid \log a_1 + \alpha'(t)\log a_2 = \beta'(t)$ and $a_1 \leq x_1(t)\}$ is higher. To the right of $x(t)$ the curve Y_t coincides with the higher of the curves $\{a \in \mathbf{E}_{++}^2 \mid \log a_1 + \alpha(t)\log a_2 = \beta(t)$ and $a_1 \geq x_1(t)\}$ and $\{a \in \mathbf{E}_{++}^2 \mid \log a_1 + \alpha'(t)\log a_2 = \beta'(t)$ and $a_1 \geq x_1(t)\}$.

Set $Y = \Pi_{t \in T} Y_t$. Then $W(Y)$ is standard by Lemma 8. Assume that t is a dictator for $f \mid Y(\alpha,\beta)$. Because $Y(\alpha,\beta) \cap Y$ contains distinct alternatives y and z such that $W_i(\{y,z\}) = P(\{y,z\})$ for all $i \in T$ individual t is a dictator for $f \mid \{y,z\}$ for any such pair. Then by Theorem 8.26 t must be a dictator for $f \mid Y$. Because $Y(\alpha',\beta') \cap Y$ contains distinct alternatives y' and z' such that $W_i(\{y',z'\}) = P(\{y',z'\})$ for all $i \in T$ individual t is a dictator for $f \mid Y(\alpha',\beta')$ by Theorem 8.26. ■

Virtually the same argument will establish the corresponding result for inverse dictatorship, which we state as Lemma 11.

11. Lemma. Suppose that $f:W \to P(\Omega)$ satisfies Arrow's independence axiom. If t is an inverse dictator for $f \mid Y(\alpha,\beta)$ and

$Y(\alpha',\beta') \cap Y(\alpha,\beta) \neq \varnothing$ then t is an inverse dictator for $f \mid Y(\alpha',\beta')$ also.

Now, suppose $Y(\alpha,\beta) \cap Y(\alpha',\beta') = \varnothing$ and t is a dictator or an inverse dictator for $f \mid Y(\alpha,\beta)$. For $i \in T$ set $Y_i = \{a \in \mathbf{E}^2_{++} \mid \log a_1 + \alpha(t)\log a_2 = \beta(t)\}$ and $Z_i = \{a \in \mathbf{E}^2_{++} \mid \log a_1 + \alpha'(t)\log a_2 = \beta'(t)\}$. If $Y_i \cap Z_i = \varnothing$ then $\alpha(i) = \alpha'(i)$ and $\beta(i) \neq \beta'(i)$. Choose any $\alpha''(i) > 0$ such that $\alpha(i) \neq \alpha''(i) \neq \alpha'(i)$. Then $Y(\alpha,\beta) \cap Y(\alpha'',\beta) \neq \varnothing$ and $Y(\alpha',\beta') \cap Y(\alpha'',\beta) \neq \varnothing$. Therefore, if t is a dictator for $f \mid Y(\alpha,\beta)$ then t is a dictator for $f \mid Y(\alpha'',\beta)$ by Lemma 10 and also for $f \mid Y(\alpha',\beta')$ by Lemma 10. If t is an inverse dictator for $f \mid Y(\alpha,\beta)$ then t is an inverse dictator for $f \mid Y(\alpha'',\beta)$ by Lemma 11 and thus for $f \mid Y(\alpha',\beta')$ by Lemma 11. Therefore, we have the following result which brings us close to proving that f itself is authoritarian.

12. Lemma. Suppose that $f : W \to P(\Omega)$ satisfies Arrow's independence axiom and $f \mid \Omega_0$ is not constant. Then there is some $t \in T$ such that t is a dictator for $f \mid Y(\alpha,\beta)$ and all $\alpha \in \mathbf{E}^{\mathrm{T}}_{++}$ and $\beta \in \mathbf{E}^{\mathrm{T}}$ or t is an inverse dictator for $f \mid Y(\alpha,\beta)$ and all $\alpha \in \mathbf{E}^{\mathrm{T}}_{++}$ and $\beta \in \mathbf{E}^{\mathrm{T}}$.

Proof. If $f \mid \Omega_0$ is not constant then by Lemma 9 $f \mid Y(\alpha,\beta)$ is authoritarian for some α and β. As we have just pointed out, this and Lemmas 10 and 11 establish the claim. ∎

All that remains is to prove that an individual who is decisive over pairs in $Y(\alpha,\beta)$ is decisive for all pairs (x,y), including cases where $x(t) \geq y(t)$ or $y(t) \geq x(t)$ for some t.

13. Theorem. A social welfare function $f : W \to P(\Omega)$ satisfies Arrow's independence axiom if and only if it is constant or completely authoritarian.

Proof. (i) It was pointed out in the first part of Theorem 8.27 that any constant or completely authoritarian social welfare function satisfies Arrow's independence axiom without restriction.

(ii) Suppose that f is a non-constant social welfare function satisfying Arrow's independence axiom. If $f \mid \Omega_0$ is constant then so is f by Lemma 4. Therefore $f \mid \Omega_0$ is not constant. Therefore, some $t \in T$ is a dictator (or an inverse dictator) for $f \mid Y(\alpha,\beta)$ and all $\alpha \in \mathbf{E}^{\mathrm{T}}_{++}$ and $\beta \in \mathbf{E}^{\mathrm{T}}$ by Lemma 12.

Suppose that t is a dictator for all $f \mid Y(\alpha,\beta)$. Choose arbitrary $x,y \in \Omega_0$ and $p \in W$. Suppose that (x,y) belongs to $Ap(t)$. If x and

y belong to $Y(\alpha,\beta)$ for some α and β then of course t dictates $f\,|\,\{x,y\}$. Otherwise, $S = \{i \in T\,|\ x(i) \geq y(i)$ implies $x(i) = y(i)$, and $x(i) \leq y(i)$ implies $x(i) = y(i)\} \neq T$. Define $Y(\alpha,\beta)$ and $Y(\alpha',\beta')$: if $i \in S$ set $\alpha(i) = \alpha'(i)$ and $\beta(i) = \beta'(i)$, where $\alpha(i)$ satisfies $\log x_1(i) + \alpha(i)x_2(i) = \log y_1(i) + \alpha(i)\log y_2(i)$ and $\beta(i) = \log x_1(i) + \alpha(i)\log x_2(i)$. If $i \notin S$ set $\alpha(i) = 1$, $\alpha'(i) = 2$, $\beta(i) = \log x_1(i) + \log x_2(i)$, and $\beta'(i) = \log y_1(i) + 2\log y_2(i)$. We have $x \in Y(\alpha,\beta)$ and $y \in Y(\alpha',\beta')$ and there is some $z \in Y(\alpha,\beta) \cap Y(\alpha',\beta')$ such that $x(i) \neq z(i) \neq y(i)\ \forall\, i \in T$. Therefore, $\{x,z\}$ and $\{z,y\}$ are free. Choose $p' \in W$ such that $(x,z),(z,y) \in Ap'(t)$ and $p'(i) = p(i)\ \forall\, i \neq t$. Then (x,z), $(z,y) \in Af(p')$ and hence $(x,y) \in Af(p')$ by transitivity. Therefore, $(x,y) \in Af(p)$ by independence and thus t is a dictator for $f\,|\,\Omega_0$.

Now, $W(\Omega_0)$ is obviously regular (8.9). Therefore, $f\,|\,\Omega_0$ is completely authoritarian by Proposition 8.10. Then $f(p) \cap \Omega_0^2 = p(t) \cap \Omega_0^2$ for all $p \in W$ and thus $f(p) = p(t)$ for all $p \in W$ by Lemma 4. That is, f is completely dictatorial.

If no individual is a dictator for all $f\,|\,Y(\alpha,\beta)$ then by Lemma 12 some individual is an inverse dictator for all $f\,|\,Y(\alpha,\beta)$. The mirror image of the above argument will show that this person is an inverse dictator for f and f is completely authoritarian. ∎

Consider now the case of quasitransitive social preferences. Theorem 8.36 carries over to the realm of welfare economics but we have to be careful in stating the theorem, as the next example demonstrates.

14. Example. $T = \{1, \ldots, n\}$ is the finite society. Set $(x,y) \in Af(p)$ if and only if $(x,y) \in Ap(1)$ *and* $x(t) \neq 0 \neq y(t)$ for all $t \in T$.

Because $(x,y) \in Sf(p)$ for $x(1) = (2,2)$, $y(1) = (1,1)$ and $x(t) = y(t) = (0,0)\ \forall\, t \neq 1$ the social welfare function f is not oligarchical or even quasi-oligarchical. Social preferences are continuous because $(x,y) \in Af(p)$ implies $x(t) \neq 0 \neq y(t)$ for all t. Clearly, strict non-imposition and Arrow's independence axiom are satisfied. Each $f\,|\,Y(\alpha,\beta)$ is oligarchical because $W(Y(\alpha,\beta))$ is standard and regular (8.36). Example 14 avoids total oligarchy by taking an oligarchical social welfare function and creating additional social indifference. This is the only logical way of avoiding total oligarchy; the next theorem proves that $f\,|\,\Omega_0$ is oligarchical if

f satisfies Arrow's independence axiom and strict non-imposition. The only acceptable oligarchical social welfare function is the Pareto aggregation rule (6.1), for which $I = T$ and $J = \varnothing$. Theorem 16 below implies that $\max R(Z)$ contains all the Pareto optimal allocations if $R = f(p)$ and $f:W \to Q(\Omega)$ satisfies strict non-imposition (3.2) and Arrow's independence axiom and $(T,\varnothing)$ is the oligarchy underlying $f \mid \Omega_0$. Therefore, if f is not in fact the Pareto aggregation rule it is even more reluctant to address distributional issues than the Pareto rule itself. This leaves us with Pareto aggregation as the only passable social welfare function that satisfies the two conditions in addition to quasitransitivity of social preference.

15. Theorem. If $f:W \to Q(\Omega)$ satisfies Arrow's independence axiom and strict non-imposition then $f \mid \Omega_0$ is completely oligarchical.

Proof. For any $\alpha \in \mathbf{E}^T_{++}$ and $\beta \in \mathbf{E}^T$ the domain $W(Y(\alpha,\beta))$ is standard (Lemma 8) and regular. Therefore, $f^1 \equiv f \mid Y(\alpha,\beta)$ is completely oligarchical by Theorem 8.36. Let (I^1,J^1) be the underlying oligarchy, and let (I^2,J^2) be the oligarchy for $f^2 \equiv f \mid Y(\alpha',\beta')$. If $Y(\alpha',\beta') \cap Y(\alpha,\beta) \neq \varnothing$ let Y be a product of SCDS curves (as in the proof of Lemma 10) such that $Y \cap Y(\alpha,\beta)$ contains a free pair $\{w,x\}$ and $Y \cap Y(\alpha',\beta')$ contains a free pair $\{y,z\}$. Because $W(Y)$ is standard $f \mid Y$ is oligarchical by Theorem 8.36. Let (I,J) be the oligarchy underlying $f \mid Y$. If $t \in I - I^1$ choose $p \in W$ so that $(x,w) \in Ap(t) \cap Ap(j)$ for all $j \in J^1$ and $(w,x) \in Ap(i)$ for all $i \in I^1$. Then $(w,x) \in Af(p)$ because (I^1,J^1) is decisive for $f \mid Y(\alpha,\beta)$, contradicting the fact that t has veto power for $f \mid Y$. Therefore, $I \subset I^1$ and a similar argument establishes $I^1 \subset I$ and $J = J^1$. Similarly, one has $I^1 = I^2$ and $J^1 = J^2$.

If $Y(\alpha',\beta') \cap Y(\alpha,\beta) = \varnothing$ choose α'',β'' such that $Y(\alpha',\beta') \cap Y(\alpha'',\beta'') \neq \varnothing \neq Y(\alpha'',\beta'') \cap Y(\alpha,\beta)$. Then (I,J) must be an oligarchy for $f \mid Y(\alpha',\beta')$. Therefore (I,J) is an oligarchy for all $f \mid Y(\alpha,\beta)$.

Suppose $(x,y) \in Ap(i) \cap -Ap(j)$ for all $i \in I$ and $j \in J$ and for arbitrary $x,y \in \Omega_0$. Choose $z \in \Omega_0$ so that $\{x,z\}$ and $\{y,z\}$ are both free and there is some $p' \in W$ such that $(x,z),(z,y) \in Ap'(i) \cap -Ap'(j)$ for all $i \in I$, $j \in J$, and $p'(t) = p(t)$ for $t \notin I \cup J$. Because $\{x,z\}$ and $\{y,z\}$ are free there exist $\alpha,\beta,\alpha',\beta'$ such that $x,z \in Y(\alpha,\beta)$ and $y,z \in Y(\alpha',\beta')$. Then $(x,z),(z,y) \in Af(p')$ and thus $(x,y) \in Af(p')$ by transitivity of $Af(p')$. Therefore, $(x,y) \in Af(p)$ by Arrow's independence axiom.

To prove that the members of I have veto power, suppose $h \in I$, $x,y \in \Omega_0$, and $(x,y) \in Ap(h) \cap -Af(p)$. Choose $\alpha,\beta,\alpha',\beta'$ such that

15a $x \in Y(\alpha,\beta)$, $y \in Y(\alpha',\beta')$.

15b $\forall i \in T$, $\alpha(i) = \alpha'(i)$, $\beta(i) = \beta'(i)$ if $x(i) \geq y(i)$ or $y(i) \geq x(i)$ implies $x(i) = y(i)$.

15c $\{x,z\}$ and $\{y,z\}$ are free for some $z \in Y(\alpha,\beta) \cap Y(\alpha',\beta')$.

15d $(x,z),(z,y) \in Ap'(h)$ for some $p' \in W$.

15e $z(i) \neq y(i) \ \forall i \in T$.

15f $W_i(\{x,y,z\} = P(\{x,y,z\})$ for any $i \in T$ such that neither $x(i) \geq y(i)$ nor $y(i) \geq x(i)$ holds.

Then we can find $p' \in W$ such that $p'(t) \cap \{x,y\}^2 = p(t) \cap \{x,y\}^2 \forall t \in T$ and (z,y) belongs to $Ap'(t)$ for all $t \in I$ and (y,z) belongs to $Ap'(t)$ for all $t \in J$. If $x(i) \geq y(i)$ or $y(i) \geq x(i)$ then any $p'(i)$ in W_i will agree with $p(i)$ over $\{x,y\}$. Otherwise 15f applies and we can find $p'(i) \in W_i$ such that $p'(i) \cap \{x,y\}^2 = p(i) \cap \{x,y\}^2$ and $(z,y) \in Ap'(i) \cap - Ap'(j)$ for all $i \in I$ and all $j \in J$ and $(x,z),(z,y) \in Ap'(h)$. Then $(z,y) \in Af(p')$ by decisiveness of (I,J) and $(y,x) \in Af(p')$ by Arrow's independence axiom. Therefore, $(z,x) \in Af(p')$ by transitivity of $Af(p')$, contradicting the fact that the members of I have veto power for $f \mid \{x,z\}$ for all free pairs. (Every free pair in Ω_0 belongs to some product of SCDS curves.) Similarly, every member of J has inverse veto power for $f \mid \Omega_0$. ∎

Let f and f' be two social welfare functions on domain D. We say that f is (weakly) finer than f' if $f(p) \subset f'(p)$ for all $p \in D$.

16. Theorem. If $f : W \to Q(\Omega)$ satisfies Arrow's independence axiom and strict non-imposition then there is a completely oligarchical social welfare function that is finer than f.

Proof. $f \mid \Omega_0$ is completely oligarchical by Theorem 15. Let (I,J) be the underlying oligarchy and define $f' : W \to Q(\Omega)$ by setting $(x,y) \in Af'(p)$ if and only if $(x,y) \in Ap(i) \cap -Ap(j)$ for all $i \in I$ and $j \in J$. Suppose that $(x,y) \in f'(p) \cap -Af(p)$. Suppose $i \in I$ and $(x,y) \in Ap(i)$. By continuity of $p(i)$ and $f(p)$ there exist neighbourhoods $N(x)$ and $N(y)$ of x and y such that

$(x',y') \in Ap(i) \cap -Af(p)$ for all $x' \in N(x)$ and $y' \in N(y)$. Choose $x' \in N(x) \cap \Omega_0$ and $y' \in N(y) \cap \Omega_0$. We have $(y',x') \in Af(p)$ although i has veto power for $f|\Omega_0$, a contradiction. Similarly, $(y,x) \in Ap(j)$ and $j \in J$ lead to a contradiction. Then we must have $(x,y) \in Sp(t)$ for some $t \in I \cup J$. But for every neighbourhood $N(x)$ of x and $N(y)$ of y there will be allocations $x',x'' \in N(x)$ and $y',y'' \in N(y)$ such that (x',y') and (y'',x'') both belong to $Ap(t)$. If $t \in I$ then $(x',y') \in Ap(t) \cap -Af(p)$ contradicts the veto power of t with respect to $f|\Omega_0$. If $t \in J$ then $(y'',x'') \in Ap(t) \cap Af(p)$ contradicts the inverse veto power of t with respect to $f|\Omega_0$. Therefore, $f'(p) \subset f(p) \; \forall p \in W$. ∎

Choose any $\omega \in \mathbf{E}_{++}^2$ and let $Z = \{x \in \Omega \,|\, \Sigma_{t \in T} x(t) \leq \omega\}$. If (I,J) is an oligarchy for f then for all $p \in W$ and $x \in \max f(p)(Z)$ we have $x(t) = (0,0)$ for some $t \notin I$. The idea that there is at least one individual who is denied any share in society's goods and services in every situation is repugnant to everyone's notion of equity. Therefore, we can say that minimal equity demands that $I = T$ and $J = \emptyset$. That is $f = f^*$, the Pareto aggregation rule. If f^* is finer than f' then $\max f^*(p)(Y)$ is a subset of $\max f'(p)(Y)$ for every profile p and every set of alternatives Y. Therefore, Pareto optimality is the only acceptable welfare standard consistent with strict non-imposition and Arrow's independence axiom. Strict non-imposition represents a substantial relaxation of the Pareto criterion. It is a simple way to formalize the notion that x is required to rank above y socially when the former leaves everyone *considerably better off* than the latter without having to specify what one means by 'considerably better off'.

11 Public Goods

SUPPOSE that a collective choice rule has been employed to determine the socially optimal mix $x(0)$ of public goods along with the allocation $x(1), x(2), \ldots, x(n)$ of private goods. If these have been arrived at by means of a social welfare function f yielding continuous and transitive social preferences and satisfying Arrow's independence axiom, then a social welfare function $f_{x(0)}$ is induced on the space Ω of allocations of private goods once a vector $x(0)$ of public goods is specified. Because $f_{x(0)}$ satisfies Arrow's independence axiom it is either constant or authoritarian by Theorem 10.13. If it is constant then individual preference and choice can play no role in the determination of the allocation of private goods once the resources required for the production of $x(0)$ have been commandeered. If $f_{x(0)}$ is authoritarian then only one person's preference scheme can have a bearing on the allocation of private goods after $x(0)$ is produced. Neither case is remotely acceptable, so Theorem 10.13, and similarly 10.15, have serious implications for economies with public goods, and we could stop generating Arrovian impossibility theorems at this point. Nevertheless, at very little extra cost we can obtain a complete extension of the results to the public goods case. First, the case of pure public goods is examined without explicitly introducing private goods into the picture and then we consider the model in which private and public goods are simultaneously featured. In every case the society is finite. Set $T = \{1, 2, \ldots, n\}$ for convenience. Only the dictatorship theorem (10.13), for fully transitive social preference, is proved. It will be obvious that the oligarchy theorem (10.15) can be extended in a like manner, although the domain assumptions would have to be strengthened slightly.

There are two public goods so the outcome space is $\mathbf{E}_+^2$ for the model with public goods only. Because everyone consumes the same vector of public goods the proofs will be relatively simple and little need be said about the domain. Our domain assumption embraces the classical case of convex, strictly monotonic, and continuous preferences employed by Kalai, *et al.* (1979) in their seminal paper, but it is more general. As usual, the domain is assumed to be a product set throughout this chapter.

1. Classical domain for $\mathbf{E}_+^2$. $D \subset P(\mathbf{E}_+^2)^\mathrm{T}$ is classical if $D(\mathbf{E}_{++}^2)$ is regular (8.9) and the following three conditions are met:

1a $D(Y)$ has the free-triple property for every strictly convex downward-sloping (SCDS) curve $Y \subset \mathbf{E}_{++}^2$.

1b   If $x,y \in \mathbf{E}_{++}^2$ and $x \geq y$ then $D_t(\{x,y\})$ is a singleton for all $t \in T$.

1c   If $x,y \in \mathbf{E}_{++}^2$ then for any $t \in T$ such that $(x,y) \in AR$ for some $R \in D_t$ there is some $R' \in D_t$ and $z \in \mathbf{E}_{++}^2$ such that (x,z), $(z,y) \in AR'$ and each of the inequalities $x \geq z$, $z \geq x$, $y \geq z$, and $z \geq y$ is false.

Regularity of $D(\mathbf{E}_{++}^2)$ is implied by weak monotonicity. If $(x,y) \in SR$ just choose $z \gg x$ sufficiently close to x to obtain $(z,x) \in AR$ with z in a neighbourhood of x. Consider 1a. If Y is an SCDS curve (10.6) then $x,y \in Y$ and $x \geq y$ implies $x = y$, so the free-triple property (3.18) is consistent with classical economic preferences when private goods are not included in the model. Obviously, 1b is implied by, but is weaker than, strict monotonicity of individual preference in the interior of the allocation space. Finally, consider 1c. If individual preferences are monotonic then 1c will be satisfied as long as each D_t contains some Cobb–Douglas preorder. Suppose that $(x,y) \in p(t)$ for some $p \in D$. Then $y \geq x$ is ruled out by monotonicity. If $x \geq y$ is also false then the alternative z and the preorder R required by 1c will exist by 1a. Suppose that $x \geq y \neq x$ holds. If α is a positive real number then R is defined by setting $(a,b) \in R$ if and only if $\log a_1 + \alpha \log a_2 \geq \log b_1 + \alpha \log b_2$. To verify 1c assume that x and y belong to $\mathbf{E}_{++}^2$ and $x \geq y \neq x$. Define $w \gg 0$ by setting $w_2 = \frac{1}{2} y_2$ and let w_1 be the solution to $\log w_1 = \log x_1 + \alpha \log x_2 - \alpha \log w_2$. Then $(x,w) \in SR$ and $(w,y) \in AR$. Because $w_2 < y_2 \leq x_2$ we must have $w_1 > x_1$ and $w_1 > y_1$. If we choose $z \ll w$ close enough to w we will have (x,z), $(z,y) \in AR$. We can also satisfy $z_2 < y_2 \leq x_2$ and $z_1 > x_1 \geq y_1$ which implies that $\{x,z\}$ is contained in some SCDS curve and that $\{z,y\}$ is also contained in some SCDS curve.

2. Theorem. If D is classical then $f: D \to P(\mathbf{E}_+^2)$ satisfies Arrow's independence axiom if and only if it is constant or completely authoritarian.

Proof: (i) A constant or completely authoritarian social welfare function satisfies Arrow's independence axiom without restriction.

(ii) Suppose that $f: D \to P(\mathbf{E}_+^2)$ satisfies Arrow's independence axiom. Suppose that f is not constant. Then $f \mid \mathbf{E}_{++}^2$ is not constant by Lemma 10.4. Then we have $(x,y) \in Af(p) \cap -f(p')$ for some $x,y \in \mathbf{E}_{++}^2$ and $p,p' \in D$. In particular, $f \mid \{x,y\}$ is not constant so 1b and Arrow's independence axiom imply that there is some SCDS curve Y containing x and y. Then $f \mid Y$ is authoritarian by property 1a and Theorem 8.8. Suppose that individual t is a dictator for $f \mid Y$.

Claim 1. If Y' is an SCDS curve and Y and Y' cross at one and only one point then t is a dictator for $f \mid Y'$.

Proof of claim 1: Suppose that Y and Y' cross at x. Define the SCDS curve $Y'' \subset \mathbf{E}_{++}^2$: To the left of x the curve Y'' coincides with whichever of the curves $\{a \in \mathbf{E}_{++}^2 \mid a \in Y$ and $a_1 \le x_1\}$ and $\{a \in \mathbf{E}_{++}^2 \mid a \in Y'$ and $a_1 \le x_1\}$ is higher. To the right of x the curve Y'' coincides with the higher of the curves $\{a \in \mathbf{E}_{++}^2 \mid a \in Y$ and $a_1 \ge x_1\}$ and $\{a \in \mathbf{E}_{++}^2 \mid a \in Y'$ and $a_1 \ge x_1\}$. Then $D(Y'')$ has the free-triple property by 1a. Because t is a dictator for $f \mid Y$ and $Y \cap Y''$ contains distinct alternatives y and z such that $D_i(\{y,z\}) = P(\{y,z\})$ for all $i \in T$ individual t is a dictator for $f \mid \{y,z\}$. Then by Theorem 8.26 t must be a dictator for $f \mid Y''$. Because $Y' \cap Y''$ contains distinct alternatives y' and z' such that $D_i(\{y',z'\}) = P(\{y',z'\})$ for all $i \in T$ individual t is a dictator for $f \mid Y'$ by Theorem 8.26. This establishes claim 1.

Virtually the same argument will establish the corresponding result for inverse dictatorship, which we state as claim 2.

Claim 2. If t is an inverse dictator for $f \mid Y$ and Y and Y' are SCDS curves that cross at one and only one point then t is an inverse dictator for $f \mid Y'$ also.

Now, suppose $Y \cap Y' = \varnothing$ and t is a dictator or an inverse dictator for $f \mid Y$. Choose an SCDS curve Y'' such that Y and Y'' cross at one and only one point and Y' and Y'' cross at one and only one point. Therefore, if t is a dictator for $f \mid Y$ then t is a dictator for $f \mid Y''$ by claim 1 and also for $f \mid Y'$ by claim 1. If t is an inverse dictator for $f \mid Y$ then t is an inverse dictator for $f \mid Y''$ by claim 2 and thus for $f \mid Y'$ by claim 2. Therefore, we have proved the following result which brings us close to proving that f itself is authoritarian.

Claim 3. There is some $t \in T$ such that t is a dictator for $f \mid Y$ for all SCDS curves Y or else t is an inverse dictator for all $f \mid Y$.

All that remains is to prove that an individual who is decisive (resp., inversely decisive) over pairs in Y is decisive (resp., inversely decisive) for all pairs (x,y), including cases where $x \geq y$ or $y \geq x$. Suppose that t is a dictator for $f \mid \{x,y\}$ for all pairs $\{x,y\}$ that are contained in some SCDS curve.

Choose arbitrary $x,y \in \mathbf{E}^2_{++}$. If neither $x \geq y$ nor $y \geq x$ holds then there is an SCDS curve Y' containing both x and y and therefore t is a dictator for $f \mid \{x,y\}$. If $x \geq y \neq x$ and $(x,y) \in Ap(t)$ then by 1c there is some $z \in \mathbf{E}^2_{++}$ and some $p' \in D$ such that $p'(i) = p(i)$ for all $i \neq t$ and $(x,z),(z,y) \in Ap'(t)$ and $\{x,z\}$ is contained in some SCDS curve, and $\{z,y\}$ is contained in some SCDS curve. Because t is a dictator for pairs that are contained in some SCDS curve we have $(x,z),(z,y) \in Af(p')$ and thus $(x,y) \in Af(p')$ by transitivity. Therefore, $(x,y) \in Af(p)$ by the independence axiom. If $y \geq x \neq y$ and $(x,y) \in Ap(t)$ then by 1c there is some $z \in \mathbf{E}^2_{++}$ and some $p' \in D$ such that $p'(i) = p(i)$ for all $i \neq t$ and $(x,z),(z,y) \in Ap'(t)$ and $\{x,z\}$ is contained in some SCDS curve, and $\{z,y\}$ is contained in some SCDS curve. Because t is a dictator for pairs that are contained in some SCDS curve we have $(x,z), (z,y) \in Af(p')$ and thus $(x,y) \in Af(p')$ by transitivity. Therefore, t is a dictator for $f \mid \mathbf{E}^2_{++}$. Because $D(\mathbf{E}^2_{++})$ is regular the social welfare function $f \mid \mathbf{E}^2_{++}$ is completely dictatorial by Proposition 8.10. Therefore, f itself is completely dictatorial by Lemma 10.4. Similarly, if $f \mid Y$ is inversely dictatorial then f is completely inversely dictatorial. ∎

Consider the complete model. Let $S = \{0,1,2, \ldots,n\}$. Agent zero is the public sector, so the allocation x assigns the commodity vector $x(t)$ to individual $t \in T$ and specifies the vector $x(0)$ of public goods. The allocation space is $\Lambda \equiv (\mathbf{E}^2_+)^S$, and Λ_0 denotes the subspace of strictly positive allocations. Now we define a classical domain of profiles in $P(\Lambda)^T$. The definition will be easier to grasp if we introduce a new binary relation $\approx_i$ on Λ. For each allocation $x,y \in \Lambda$ and any $i \in S$ set $x \approx_i y$ if both $x(i) \geq y(i)$ and $y(i) \geq x(i)$ are false. Note that $x \approx_i y$ implies that $x(i)$ and $y(i)$

belong to an SCDS curve (10.6) if $x(i) \geqslant 0$ and $y(i) \geqslant 0$. As in the other chapters, the domain is assumed to be a product set. (Recall the definition of F_t prior to 8.15.)

3. Classical domain for Λ. The domain $D \subset P(\Lambda)^T$ is classical if it is regular (8.9) and the following six conditions are met for arbitrary $t \in T, x, y, z \in \Lambda_0$, and $Y, Z \subset \Lambda_0$.

3a   If $x \notin F_t(y)$ then $D_t(\{x,y\})$ is a singleton.

3b $D_t(\{x,y,z\}) = P(\{x,y,x\})$ if $y \in F_t(x) \cap F_t(z)$ and for $i \in \{0,t\}$:

3b.1 $x \approx_i y$ implies that $x(i), y(i)$, and $z(i)$ are distinct members of some SCDS curve,

3b.2 $x(i) \geq y(i)$ implies $z(i) \geq y(i)$ and either $x(i) \geq z(i)$ or $x \approx_i z$,

3b.3 $x(i) \leq y(i)$ implies $z(i) \leq y(i)$ and either $x(i) \leq z(i)$ or $x \approx_i z$.

3c $F(x) \cap F(y) \neq \emptyset$.

3d $D(Y)$ is standard if Y is a product of SCDS curves.

3e   If Y and Z are products of SCDS curves and $Y \cap Z$ is uncountable then $Y \cap Z$ contains a free pair.

3f   If Y and Z are intersecting products of SCDS curves and $(a,b) \in AR$ for some $a \in Y$, $b \in Z$, and $R \in D_t$ then there exist $c \in Y \cap Z$ and $R' \in D_t$ such that $(a,c), (c,b) \in AR'$.

To motivate condition 3a consider two allocations x and y such that $x(1)$ provides individual 1 with far more of each private good than $y(1)$ but $x \approx_0 y$, although $x(0)$ and $y(0)$ are close in Euclidean distance. It would be rash to assume that D_1 contains a preorder that ranks y above x, and we allow the possibility that every member of D_1 ranks x strictly above y. But if D_1 contains two preorders that disagree with respect to the pair $\{x,y\}$ then $D_t(\{x,y\})$ must contain all three logically possible preorders on $\{x,y\}$. To motivate 3b assume that individual t cares only about his own consumption of private goods and the vector of public goods provided. $D_t(\{x,y,z\}) = P(\{x,y,z\})$ would be expected to hold if $x(i), y(i)$, and $z(i)$ were distinct members of some SCDS curve for $i \in \{0,t\}$. The set $\{x(i), y(i), z(i)\}$ can be arbitrarily ordered by means of an appropriately chosen classical indifference map and the desired member of $P(\{x,y,z\})$ can be generated by putting sufficient weight on the consumption of private goods by individual t.

But suppose that we have $x(0) \gg z(0) \gg y(0)$. If $x \approx_t y$ does not hold then we can rule out $x(t) \geq y(t)$ because y belongs to $F_t(x)$. If $y(t) \geq x(t)$ then $y(t) \geq z(t) \geq x(t)$ by 3b.3, in addition to $x(0) \gg z(0) \gg y(0)$. In that case we would expect $D_t(\{x,y,z\}) = P(\{x,y,z\})$ to hold if neither public goods nor private goods *always* dominated individual t's preferences. Suppose that $x(0) \gg z(0) \gg y(0)$ holds and $x \approx_t y$ or $z \approx_t y$ also holds. Then $x(t)$, $y(t)$, and $z(t)$ belong to some SCDS curve under the hypothesis of 3b. Because y belongs to $F_t(x)$ and $x(0) \gg y(0)$ we may assume that D_t contains preferences with respect to which public goods do not dominate, and we can call upon an arbitrary ordering of x, y, and z by placing sufficient weight on the private goods component of t's consumption. Condition 3c can be expected to hold because for arbitrary allocations x and y we can find an allocation z such that x and z belong to some product of SCDS curves and y and z belong to another such product. Condition 3d obviously allows us to take advantage of Theorem 8.26. It is a modest assumption, and 3e is extremely mild. Condition 3f requires some explanation. If Y and Z are products of SCDS curves their intersection will typically be a singleton, say $Y \cap Z = \{c\}$. Suppose that a belongs to Y, b belongs to Z and $a(t) \geq b(t) \neq a(t)$. Set $x(t) = \frac{1}{2}a(t) + \frac{1}{2}b(t)$. If $c(t) \geq x(t)$ then $c(t) \geq b(t) \neq c(t)$ and as a result $b(t)$ and $c(t)$ cannot belong to the same SCDS curve. If $c(t) \leq x(t)$ then $a(t)$ and $c(t)$ cannot belong to the same SCDS curve. Therefore, $c_1(t) > x_1(t)$ implies $x_2(t) > c_2(t)$ and $c_1(t) < x_1(t)$ implies $c_2(t) > x_2(t)$. Therefore, the straight line connecting $c(t)$ and $x(t)$ has negative slope. Let α_1 and α_2 be the coefficients of that line and define the linear function u on $\mathbf{E}_+^2$ by setting $u(y_1,y_2) = \alpha_1 y_1 + \alpha_2 y_2$. Then $u(a(t)) > u(c(t)) > u(b(t))$. Given the hypothesis of 3f we can generate the desired ranking $(a,c),(c,b) \in AR'$ by placing sufficient weight on the private components of t's consumption. If $b(0) \gg a(0)$ and b offers far more of some public goods than a the fact that (a,b) belongs to AR for some $R \in D_t$ indicates there are preference schemes under which t's private consumption can dominate.

Two lemmas precede the theorem.

4. Lemma. If $f: D \to P(\Lambda)$ is a non-constant social welfare function satisfying Arrow's independence axiom and $D \subset P(\Lambda)^{\mathrm{T}}$ is classical then there exists $x,y \in D$ and $p,p' \in \Lambda_0$ such that $(x,y) \in Af(p) \cap -Af(p')$.

Proof. $f \mid \Lambda_0$ is not constant by Lemma 10.4. Therefore, we have $(x,y) \in Af(p) \cap -f(p')$ for some $x,y \in \Lambda_0$ and $p,p' \in D$. Define the subspace Y of Λ_0. For $t \in S$, if $x \approx_t y$ set $Y_t = \{a \in \mathbf{E}^2_{++} \mid \log a_1 + \alpha(t)\log a_2 = \log x_1(t) + \alpha(t)\log x_2(t)\}$, where $\alpha(t) = [\log x_1(t) - \log y_1(t)] / [\log y_2(t) - \log x_2(t)]$. Otherwise, set $Y_t = \{\lambda x(t) + (1-\lambda)y(t) \mid 0 \le \lambda \le 1\}$. Each Y_t is connected by Propositions 2.13 and 2.14. Therefore, $Y = \Pi_{t \in S} Y_t$ is connected by Proposition 2.18. Allocations x and y belong to Y by construction. Therefore, there exists some $z \in Y$ such that $(x,z),(z,y) \in Af(p)$ by Proposition 2.19. We can choose z so that, for all $t \in S$, $y(t) \ne z(t) \ne x(t)$ if $x(t) \ne y(t)$. Now we locate a profile $p'' \in D$ such that, for each $t \in T$, $p''(t) \cap \{z,y\}^2 = p(t) \cap \{z,y\}^2$ and $p''(t) \cap \{y,x\}^2 = p'(t) \cap \{y,x\}^2$. If $y \notin F_t(z)$ set $p''(t) = p'(t)$ and if $y \notin F_t(x)$ set $p''(t) = p(t)$. Then $p''(t)$ has the desired property by 3a. If $y \in F_t(z) \cap F_t(x)$ then the desired preorder $p''(t)$ exists by 3b. Then $(z,y) \in Af(p'')$ and $(y,x) \in f(p'')$ by the independence axiom. Therefore, $(z,x) \in Af(p'')$ by transitivity of $f(p'')$. We already have $(x,z) \in Af(p)$. ∎

As in Chapter 10, for real numbers $\alpha(t) > 0$ and $\beta(t)$, $\forall t \in S$, we define the associated product of SCDS curves:

$$Y(\alpha,\beta) = \{x \in \Lambda_0 \mid \log x_1(t) + \alpha(t)\log x_2(t) = \beta(t) \, \forall t \in S\}.$$

5. Lemma. If $f: D \to P(\Lambda)$ is a non-constant social welfare function satisfying Arrow's independence axiom and $D \subset P(\Lambda)^T$ is classical then there exist α,β such that $f \mid Y(\alpha,\beta)$ is not constant.

Proof. By Lemma 4 we have $(x,y) \in Af(p) \cap -Af(p')$ for some $p,p' \in D$. By continuity of social preference we can assume that $x(i) \ne y(i) \, \forall i \in S$. Then there is a neighbourhood $N(x)$ of x such that $(z,y) \in Af(p) \cap -Af(p')$ for all $z \in N(x)$. Define $Y(\alpha,\beta)$ by using $\alpha(t)$ and $\beta(t)$ of Lemma 4 if $x \approx_t y$. Otherwise, set $\alpha(t) = 1$ and $\beta(t) = \log x_1(t) + \log x_2(t)$. Now choose $z \in N(x) \cap Y(\alpha,\beta)$ so that $x(t) \ne z(t)$, and $x(t) \ge y(t)$ implies $z(t) \ge y(t)$, and $x(t) \le y(t)$ implies $z(t) \le y(t)$, $\forall t \in S$. Now locate two profiles p^1 and p^2 in D. If $y \notin F_t(x)$ set $p^1(t) = p(t)$ and $p^2(t) = p'(t)$. If $y \notin F_t(z)$ set $p^1(t) = p'(t)$ and $p^2(t) = p(t)$. In either case we have $p^1(t) \cap \{z,y\}^2 = p(t) \cap \{z,y\}^2$, $p^1(t) \cap \{y,x\}^2 = p'(t) \cap \{y,x\}^2$, $p^2(t) \cap \{z,y\}^2 = p'(t) \cap \{z,y\}^2$, and $p^2(t) \cap \{y,x\}^2 = p(t) \cap \{y,x\}^2$ by 3a. And we can establish these equalities if $y \in F_t(x) \cap F_t(z)$ by virtue of 3b. Therefore, $(z,y),(y,x) \in Af(p^1) \cap -Af(p^2)$ by Arrow's independence axiom. Therefore, $(z,x) \in Af(p^1) \cap -Af(p^2)$ by transitivity and hence $f \mid Y(\alpha,\beta)$ is not constant. ∎

6. Theorem. Assume a classical domain $D \subset P(\Lambda)^T$ and a finite society T. A social welfare function $f: D \to P(\Lambda)$ satisfies Arrow's independence axiom if and only if it is constant or completely authoritarian.

Proof. (i) A constant or completely authoritarian social welfare function satisfies Arrow's independence axiom without restriction.

(ii) By Lemma 5, $f \mid Y(\alpha,\beta)$ is not constant for some choice of α and β if f is not constant. By 3d and Theorem 8.26 the social welfare function $f \mid Y(\alpha,\beta)$ is authoritarian.

Claim 1. If individual 1 is a dictator for $f \mid Y(\alpha,\beta)$ and $Y(\alpha',\beta') \cap Y(\alpha,\beta) \neq \varnothing$ then 1 is a dictator for $f \mid Y(\alpha',\beta')$.

Proof. Suppose $x \in Y(\alpha',\beta') \cap Y(\alpha,\beta)$. For each $t \in S$ define the SCDS curve $Y_t \subset \mathbf{E}_{++}^2$ as follows.

To the left of $x(t)$ the curve Y_t coincides with whichever of the curves $\{a \in \mathbf{E}_{++}^2 \mid \log a_1 + \alpha(t)\log a_2 = \beta(t) \text{ and } a_1 \leq x_1(t)\}$ and $\{a \in \mathbf{E}_{++}^2 \mid \log a_1 + \alpha'(t)\log a_2 = \beta'(t) \text{ and } a_1 \leq x_1(t)\}$ is higher. To the right of $x(t)$ the curve Y_t coincides with the higher of the curves $\{a \in \mathbf{E}_{++}^2 \mid \log a_1 + \alpha(t)\log a_2 = \beta(t) \text{ and } a_1 \geq x_1(t)\}$ and $\{a \in \mathbf{E}_{++}^2 \mid \log a_1 + \alpha'(t)\log a_2 = \beta'(t) \text{ and } a_1 \geq x_1(t)\}$.

Set $Y = \Pi_{t \in S} Y_t$. Then $D(Y)$ is standard by 3d. Because 1 is a dictator for $f \mid Y(\alpha,\beta)$ and $Y(\alpha,\beta) \cap Y$ contains distinct alternatives y and z such that $D_t(\{y,z\}) = P(\{y,z\})$ for all $t \in T$ individual 1 is a dictator for $f \mid \{y,z\}$. (Use 3e.) Then by Theorem 8.26 person 1 must be a dictator for $f \mid Y$. By 3e the set $Y(\alpha',\beta') \cap Y$ contains distinct alternatives y' and z' such that $D_t(\{y',z'\}) = P(\{y',z'\})$ for all $t \in t$, and therefore individual 1 is a dictator for $f \mid Y(\alpha',\beta')$ by Theorem 8.26.

Virtually the same argument will establish the corresponding result for inverse dictatorship, which we state as claim 2.

Claim 2. If 1 is an inverse dictator for $f \mid Y(\alpha,\beta)$ and $Y(\alpha',\beta') \cap Y(\alpha,\beta) \neq \varnothing$ then 1 is an inverse dictator for $f \mid Y(\alpha',\beta')$ also.

Now, suppose $Y(\alpha,\beta) \cap Y(\alpha',\beta') = \varnothing$ and 1 is a dictator or an inverse dictator for $f \mid Y(\alpha,\beta)$. For $t \in S$ set $Y_t = \{a \in \mathbf{E}^2_{++} \mid \log a_1 + \alpha(t)\log a_2 = \beta(t)\}$ and $Z_t = \{a \in \mathbf{E}^2_{++} \mid \log a_1 + \alpha'(t)\log a_2 = |\beta'(t)|\}$. If $Y_t \cap Z_t = \varnothing$ then $\alpha(t) = \alpha'(t)$ and $\beta(t) \neq \beta(t)$. Choose any $\alpha''(t) > 0$ such that $\alpha(t) \neq \alpha''(t) \neq \alpha'(t)$. Then $Y(\alpha,\beta) \cap Y(\alpha'',\beta) \neq \varnothing$ and $Y(\alpha',\beta') \cap Y(\alpha'',\beta) \neq \varnothing$. Therefore, if 1 is a dictator for $f \mid Y(\alpha,\beta)$ then 1 is a dictator for $f \mid Y(\alpha'',\beta)$ by claim 1 and also for $f \mid Y(\alpha',\beta')$ by claim 1. If 1 is an inverse dictator for $f \mid Y(\alpha,\beta)$ then 1 is an inverse dictator for $f \mid Y(\alpha'',\beta)$ by claim 2 and thus for $f \mid Y(\alpha',\beta')$ by claim 2. Therefore, we have the following result which brings us close to proving that f itself is authoritarian.

Claim 3. There is some $t \in T$ such that t is a dictator for $f \mid Y(\alpha,\beta)$ and all α,β or t is an inverse dictator for $f \mid Y(\alpha,\beta)$ and all α,β.

All that remains is to prove that an individual who is decisive (resp., inversely decisive) over pairs in $Y(\alpha,\beta)$ is decisive (resp., inversely decisive) for all pairs (x,y), including cases where $x(t) \geq y(t)$ or $y(t) \geq x(t)$ for some $t \in S$.

Suppose that t is a dictator for all $f \mid Y(\alpha,\beta)$ and $x,y \in \Lambda_0$, $p \in D$, and $(x,y) \in Ap(t)$. If x and y belong to $Y(\alpha,\beta)$ for some α and β then of course t dictates $f \mid \{x,y\}$. Otherwise, $K = \{i \in S \mid x \approx_i y$ or $x(i) = y(i)\} \neq S$. Define $Y(\alpha,\beta)$ and $Y(\alpha',\beta')$. If $i \in K$ set $\alpha(i) = \alpha'(i)$ and $\beta(i) = \beta'(i)$, where $\alpha(i)$ satisfies

$$\log x_1(i) + \alpha(i)x_2(i) = \log y_1(i) + \alpha(i)\log y_2(i)$$

and $\beta(i) = \log x_1(i) + \alpha(i)\log x_2(i)$.

If $i \notin K$ set $\alpha(i) = 1$, $\alpha'(i) = 2$, $\beta(i) = \log x_1(i) + \log x_2(i)$, and $\beta'(i) = \log y_1(i) + 2\log y_2(i)$. We have $x \in Y(\alpha,\beta)$ and $y \in Y(\alpha',\beta')$ with $Y(\alpha,\beta) \cap Y(\alpha',\beta') \neq \varnothing$. Therefore, by 3f we can find $z \in Y(\alpha,\beta) \cap Y(\alpha',\beta')$ and $p' \in D$ such that $(x,z),(z,y) \in Ap'(t)$ and $p'(i) = p(i) \, \forall i \neq t$. Then $(x,z),(z,y) \in Af(p')$ because t dictates $f \mid Y(\alpha,\beta)$ and $f \mid Y(\alpha',\beta')$. Then $(x,y) \in Af(p)$ by transitivity and Arrow's independence axiom. Therefore, t dictates $f \mid \Lambda_0$ and thus t is a complete dictator because $D(\Lambda_0)$ is regular (8.10). Therefore t is a complete dictator for f by Lemma 10.4. Similarly, if $f \mid Y(\alpha,\beta)$ is inversely dictatorial for some α and β then f is completely inversely dictatorial. ∎

12 Overlapping Generations

THE overlapping generations model introduced by P. A. Samuelson is very useful for studying public policy in a dynamic setting (Samuelson 1958). An early paper in this vein is Diamond (1970) on tax incidence; the book by Auerbach and Kotlikoff (1987) on dynamic fiscal policy is a more recent example. The overlapping generations model often exhibits multiple competitive equilibria and non-trivial inefficiencies, and its properties are quite different from those of competitive equilibrium models with a finite number of commodities and traders. Therefore, it cannot be taken for granted that social choice theorems will carry over to the overlapping generations regime. This chapter demonstrates that the main theorems of this book do indeed go through. In fact, they are confirmed even when the space of alternatives is the set of *feasible* allocations. The feasible set is not a product set, so the proofs that work for the entire allocation space do not directly apply. They are employed in this chapter as intermediate products, however. Only the pure private goods case will be examined, and this time we begin with the case of quasitransitive social preference. Recall the definition of an elementary domain (9.8).

Once again, $\Omega \equiv (\mathbf{E}_+^2)^T$ denotes the standard set of allocations of two private goods over the set of agents T, and $\Omega_0 \equiv (\mathbf{E}_{++}^2)^T$ is the set of interior allocations. Then $x \in \Omega$ assigns $x_c(t)$ units of commodity c to agent t. And $W \subset P(\Omega)^T$ represents the domain of profiles p such that each $p(t)$ complete, transitive, continuous, selfish, monotonic, and convex. Of course, W is not elementary because it is not standard (9.8). As in Chapter 10 we will apply the main theorem (9.13 in this case) to products of strictly convex downward-sloping curves and then work our way to the entire space of allocations, and eventually to the subspace of *feasible* allocations. It is not necessary to assume the product topology for Ω. The theorem will go through with a much weaker topology. (Continuity of social preference with respect to the product topology is very demanding when T is infinite.)

Because Ω is a product set our first oligarchy theorem for economic environments, Theorem 1, does not apply directly to overlapping generations economies. Feasible sets cannot be product sets in resource allocation models. In the case of the overlapping generations model an *endowment* is a member ω of $(\mathbf{E}^2_{++})^T$, and we say that allocation $x \in \Omega$ is *feasible* if and only if $x_1(1) \le \omega_1(1)$ and $x_2(t) + x_1(t+1) \le \omega_2(t) + \omega_1(t+1)$ for $t = 1,2,\ldots$. (Each generation lives for two periods, and in its second period of life generation t coexists with generation $t + 1$ in its initial period of life.) Let B denote the set of feasible consumption streams when ω specifies the endowments. We will show that f is completely oligarchical when $f: W(B) \to Q(B)$ satisfies Arrow's independence axiom and strict non-imposition. The corresponding theorem for outcome set Ω and domain W will be used to establish a crucial lemma, so we investigate that case first.

We begin by recalling 10.6, the definition of a strictly convex downward-sloping (SCDS) curve in $\mathbf{E}^2_{++}$, the strictly positive quadrant. $C \subset \mathbf{E}^2_{++}$ is an SCDS curve if C is arc-connected and for all distinct $x,y,z \in C$,

$$(x_1 - y_1)(x_2 - y_2) < 0 \text{ and}$$
$$x_1 < y_1 < z_1 \text{ implies } (y_2 - x_2)/(y_1 - x_1) < (z_2 - y_2)/(z_1 - y_1).$$

We also required that for arbitrary $n > 0$ there exist $v,w \in C$ such that $v_1 > n$ and $w_2 > n$. This last property is not essential so we dispense with it here. (The curves are constructed in Chapter 10 so that they have this third property. This makes it easier to construct a classical indifference curve by perturbing an SCDS curve, and that is the only role played by the third property.)

Recall that the set of strictly positive allocations is denoted $\Omega_0 \equiv (\mathbf{E}^2_{++})^T$. Before presenting the formal proofs we need to consider the choice of a topology for Ω. We want a fairly weak topology—to make it easier to find continuous social preferences—but we also want a meaningful topology so that the social preference ordering has some practical value. The second consideration demands a topology that is not error prone. Suppose that x is preferred to y socially and x' is close to x in some practical sense. For example, for each t in T the vector $x(t)$ could be so close to $x'(t)$ in Euclidean distance that the two commodity vectors are practically indistinguishable. Then x' should be socially preferred to y for the original scheme of individual preferences. As a minimal step toward satisfying this requirement we assume that (i) if x belongs to some open set N then there is some $N_t \subset \mathbf{E}_+^2$ such that N_t is open in the Euclidean topology and $y \in N$ whenever $y(t) \in N_t$ and $y(i) = x(i) \; \forall \, i \neq t$. We also require the following: if $Y = \Pi_{t \in \mathrm{T}} Y_t$ is the product of SCDS curves then (ii) Y is connected in the topology on Ω, and (iii), there exist distinct $x, y \in Y$ and associated sequences $\{x^n\}$, $\{y^n\}$ converging to x and y respectively and satisfying

$$x_1^n(t) < x_1^{n+1}(t) < x_1(t) < y_1(t) < y_1^{n+1}(t) < y_1^n(t)$$

for all $t \in T$ and every positive integer n. Throughout this chapter properties (i)–(iii) are implicit. They are exhibited by the 'sup' topology on Ω for example. (The members of W_t are still assumed to be continuous with respect to the product topology. That is, we do not allow the choice of topology for Ω to affect the definition of W.) Say that f is *strictly* oligarchical if the oligarchy is finite.

1. Lemma. If $Y = \Pi_{t \in \mathrm{T}} Y_t$ is a product of SCDS curves then $W(Y)$ is elementary.

Proof. $W(Y)$ is standard by *Lemma* 10.8.

We need to show that the following sequences required for 8a–8c of Definition 9.8 can be obtained.

8a $\quad x^i \neq x^j \neq x$ and $y^i \neq y^j \neq y$ if $i \neq j$.

8b $\quad \{x^n, y^n\}$ is free for all n.

8c $\quad$ For each $t \in T$ there exist preorders $R^1(t), R^2(t), R^3(t), R^4(t)$, and $R^5(t)$ in D_t with the following properties:

$\quad (x, y^n), (x, y), (x^n, y^n), (x^n, y) \in AR^1(t) \; \forall \, n \in T.$

$\quad (y^n, x), (y, x), (y^n, x^n), (y, x^n) \in AR^2(t) \; \forall \, n \in T.$

$\quad (x^n, x), (y, y^n) \in AR^3(t) \; \forall \, n \in T$ and $(x, y) \in SR^3(t).$

$\quad (x, x^n), (x, y), (y^m, x^n), (y^m, y) \in AR^4(t)$ if $m \leq t$ and $n \leq t.$

$\quad (x^n, y^m) \in AR^4(t)$ if $m > t$ and $n > t.$

$\quad (x^n, x), (y, x), (x^n, y^m), (y, y^m) \in AR^5(t)$ if $m \leqslant t$ and $n \leqslant t.$

$\quad (y^m, x^n) \in AR^5(t)$ if $m > t$ and $n > t.$

First choose $x(t)$ and $y(t)$ in Y_t such that $x_1(t) < y_1(t)$. The pair $\{x, y\}$ is free by Lemma 10.7. Let $\{x^n(t)\}$ and $\{y^n(t)\}$ be any sequences in Y_t converging to x and y with the additional property that $x_1^n(t) < x_1^{n+1}(t) < x_1(t)$ and $y_1(t) < y_1^{n+1}(t) < y_1^n(t)$ for all n. Then it is easy to find a strictly positive vector $\alpha = (\alpha_1, \alpha_2) \in \mathbf{E}_+^2$ such that $R^1(t)$ has the desired property when we set $(w, z) \in R^1(t)$ if and only if $\alpha_1 w_1(t) + \alpha_2 w_2(t) \geq \alpha_1 z_1(t) + \alpha_2 z_2(t)$. Just let $\alpha_1 > 0$ and $\alpha_2 > 0$ be the coefficients of a straight line cutting through Y_t so that x and x^n lie above the line for all n and y and y^n lie below the line for all n. $R^1(t)$ obviously belongs to W_t. Similarly, it is easy to define $R^2(t)$ as desired.

To define $R^3(t)$ begin by choosing any $z(t) \in Y_t$ such that $z_1(t) > y_1^1(t)$. The indifference curve C through $x(t)$ and $y(t)$ is composed of two line segments meeting at $y(t)$. The left-hand piece is just the line through $x(t)$ and $y(t)$ truncated at $y(t)$ on the right and the vertical axis on the left. The right-hand piece is the line through $y(t)$ and $z(t)$ truncated at $y(t)$ on the left and at the horizontal axis on the right. Define $u: \mathrm{E}_+^2 \to \mathrm{E}^1$ as in Lemma 10.7. Let $\alpha(w)$ be the positive real number such that $\alpha(w)w$ belongs to C. Set $(w,z) \in R^3(t)$ if and only if $u(w(t)) \geq u(z(t))$. We have $R^3(t) \in W_t$.

Define $R^4(t)$ in a similar fashion. We merely specify an indifference curve C through $x(t)$ and use it to define a utility function on E_+^2 that will give us $R^4(t)$. C is composed of four line segments. To define these, begin by choosing $\epsilon > 0$, v on Y_t to the left of $x^1(t)$, and z on Y_t between $x(t)$ and $y(t)$. The leftmost segment of the indifference curve is the line segment through v and $x^{t+1}(t)$ ending at the vertical axis on the left and the point w^ϵ to the right, where w^ϵ is that point on the line through v and $x^{t+1}(t)$ that is a distance of ϵ to the right of $x^{t+1}(t)$. The second line segment passes through w^ϵ and $x(t)$ and has those two points as beginning and end. The third segment begins at $x(t)$ and ends at z. The last segment begins at z, passes through $y'(t)$ and ends at the horizontal axis. For $\epsilon > 0$ sufficiently small each segment of the indifference curve has a slope that is smaller in absolute value than the piece meeting it to the left. $R^5(t)$ is defined in the analogous way. ■

2. Lemma. If $Y = \Pi_{t \in \mathrm{T}} Y_t$ is a product of SCDS curves then $W(Y)$ is basic.

Proof. Recall Definition 9.1 of a basic domain. $W(Y)$ is standard by Lemma 10.8. Let x and y be any two members of Y. We may assume that $x_1(t) \leq y_1(t)$. If $x_1(t) < y_1(t)$ then define $R^4(t)$ as in the proof of Lemma 1. If $x_1(t) = y_1(t)$ the obvious modification of this procedure will complete the proof. ■

3. Theorem. If $f: W(\Omega_0) \to Q(\Omega_0)$ satisfies strict non-imposition and Arrow's independence axiom then it is strictly oligarchical.

Proof. Let $Y_t^i = \{x \in E_{++}^2 \,|\, \log x_1 + i\log x_2 = 1\}$ and set $Y^i = \Pi_{t\in T} Y_t^i, i = 1,2$. By lemma 1 and Theorem 9.13, $f^i \equiv f \,|\, W(Y^i)$ is oligarchical for $i = 1,2$. Let (I^i, J^i) be the minimal decisive pair underlying f^i. Both I^i and J^i are finite by 9.13. Now define Y_t^3 and hence $Y^3 = \Pi_{t\in T} Y_t^3$: Let $x^*(t) = (e,1)$ be the intersection of Y_t^1 and Y_t^2. By the implicit function theorem of calculus the slope of Y_t^1 at $a = (a_1, a_2)$ is $-a_2/a_1$ and the slope of Y_t^2 at a is $-a_2/2a_1$. Therefore, Y_t^2 is flatter than Y_t^1 at $(e,1)$ and it lies below Y_t^1 to the left of $(e, 1)$ and above Y_t^1 to the right of $(e,1)$. Let Y_t^3 coincide with Y_t^1 to the left of $x^*(t)$ and with Y_t^2 to the right of $x^*(t)$. Because Y_t^3 is a an SCDS curve the social welfare function $f^3 \equiv f \,|\, W(Y^3)$ is oligarchical by Lemma 1 and Theorem 9.13. Because $W(Y^3)$ is elementary and Y^3 and Y^i have pairs of free points in common the oligarchies for f^3 and f^i are identical. The pair (I^1, J^1) is decisive for free pairs belonging to $Y^3 \cap Y^1$. And if (I^3, J^3) is an oligarchy for f^3 we must have $I^1 \subset I^3$ because members of I^1 have veto power over all pairs in $Y^3 \cap Y^1$. We have $J^1 \subset J^3$ because the members of J^1 have inverse veto power over all pairs in $Y^3 \subset Y^1$. But we also have $I^3 \subset I^1$ and $J^3 \subset J^1$ because the members of I^3 and J^3 have veto power and inverse veto power, respectively. Similarly, $I^2 = I^3$ and $J^2 = J^3$. Set $I^* = I^3$ and $J^* = J^3$. It is an easy matter to extend this argument to prove that (I^*, J^*) is an oligarchy for $f \,|\, Y$ for any product Y of SCDS curves. (If $Y \cap Y^1 = \varnothing$ choose Y' such that $Y \cap Y' \neq \varnothing$ and $Y' \cap Y^1 \neq \varnothing$.)

Suppose that x and y belong to Ω_0 and $\{x,y\}$ is free. Then neither $x(t) \geq y(t)$ nor $y(t) \geq x(t)$ holds for any $t \in T$ and therefore there is a product of SCDS curves containing x and y. By the above argument (I^*, J^*) is an oligarchy for $f \,|\, \{x,y\}$. Suppose x and y belong to Ω_0 and $(x,y) \in \cap_{i\in I} Ap(i) \cap_{j\in J} - Ap(j)$ but there is no product of SCDS curves containing both x and y. We can find $z \in \Omega_0$ such that $\{x,z\}$ and $\{y,z\}$ are free and there is a profile $p' \in W$ such that $(x,z),(z,y)$ belongs to $\cap_{i\in I} Ap(i) \cap_{j\in J} - Ap(j)$. (Choose two intersecting products Y and Z of SCDS curves meeting at, say, z.) Then $(x,z),(z,y) \in Af(p')$ by the above argument, and hence $(x,y) \in Af(p') \cap Af(p)$ by transitivity and the independence axiom. The last part of the proof of Theorem 10.15 shows that the members of a minimal decisive pair of coalitions (I^*, J^*) for $f: W(\Omega_0) \to Q(\Omega_0)$ have veto power (if they belong to I^*) and inverse veto power (if they belong to J^*). The argument is valid for any society T, although a different route was taken in Chapter 10 to prove the existence of a minimal decisive pair for a finite society. ■

Both the 'sup' and product topologies imply that f is *completely* oligarchical under the hypothesis of Theorem 3. In fact, this will be the case with any topology for which every neighbourhood of an arbitrary point x contains a point z such that $z(t) \gg x(t)$. Suppose this to be the case but $(y,x) \in Af(p) \cap Sp(t)$ and $t \in I^*$. If $N(x)$ is a neighbourhood of x then we will have $z(t) \gg x(t)$ for some $z \in N(x)$. By selfishness and monotonicity we have $(z,x) \in Ap(t)$. Because $(x,y) \in Sp(t)$ we have $(z,y) \in Ap(t)$ and thus, by veto power of t, $(z,y) \in f(p)$. Therefore, $f(p)$ is not continuous. Similarly, if (y,x) belongs to $Af(p) \cap Sp(t)$ and $t \in J^*$ choose z arbitrarily close to y such that $z(t) \gg y(t)$. Then $(z,x) \in Ap(t)$ and thus $(x,z) \in f(p)$ by inverse veto power of t. This also contradicts the continuity of $f(p)$. In other words, continuity of social preference implies that the oligarchy is complete.

Now we fix the endowment point ω for the remainder of the chapter and turn our attention to the space B of feasible allocations and the domain $W(B)$. Set $\omega_2(0) = 0$. Then $B = \{x \in \Omega \mid x_2(t-1) + x_1(t) \le \omega_2(t-1) + \omega_1(t) \ \forall \, t \in T\}$ and we let B_0 denote the set $\{x \in \Omega \mid (0,0) \ll x_2(t-1) + x_1(t) \ll \omega_2(t-1) + \omega_1(t) \ \forall \, t \in T\}$. We begin with a generalization of Theorem 3. First, choose $x \in \Omega_0$ and set $Z(x) = \{y \in \Omega_0 \mid y(t) \ll x(t) \ \forall \, t \in T\}$.

4. Lemma. If $f: W(B_0) \to Q(B_0)$ satisfies strict non-imposition and Arrow's independence axiom then $f \mid Z(x)$ is strictly oligarchical if $Z(x) \subset B_0$.

Proof. $Z(x)$ is a product set and if Y is the product of SCDS curves then $W(Z(x) \cap Y)$ is elementary. Both Y and $Z(x) \cap Y$ are products of curves in $\mathbf{E}_+^2$ that are homoeomorphic to an open interval in the real line. The fact that latter is bounded is no obstacle to the proof of Theorem 3 which can be employed here without modification. ■

Because $Z(\omega)$ is a subset of B_0 we can apply Lemma 4 to $f \mid Z(\omega)$. Let (I,J) be the oligarchy for $f \mid Z(\omega)$. We will show that (I,J) is an oligarchy for f.

5. Lemma. If $f: W(B_0) \to Q(B_0)$ satisfies strict non-imposition and Arrow's independence axiom then $f \mid \{x,y\}$ is oligarchical for any $x, y \in B_0$ such that $x \in F_t(y) \; \forall \, t \in I \cup J$, and (I, J) is the underlying oligarchy.

Proof. Let x and y be any two members of B_0. Choose a and b in B_0 such that $x \in Z(a)$ and $y \in Z(b)$. To show that this is possible let $\varepsilon(t) = \omega_1(t) + \omega_2(t - 1) - x_1(t) - x_2(t - 1)$ and let $\delta(t) = \frac{1}{2}\min\{\varepsilon(t), \varepsilon(t + 1)\}$. We can set $a(t) = x(t) + (\delta(t), \delta(t))$. This defines a member of B_0. Now, define $e \in B_0$ by setting $e_c = \min\{a_c, b_c\}$ for $c = 1, 2$ and $\forall \, t \in T$. Suppose that x belongs to $F_t(y) \; \forall \, t \in I \cup J$. Then for any $t \in I \cup J$ either $x_1(t) > y_1(t)$ and $x_2(t) < y_2(t)$ both hold, or else $x_1(t) < y_1(t)$ and $x_2(t) > y_2(t)$ both hold. If $x_1(t) > y_1(t)$ and $x_2(t) < y_2(t)$ then we can assume that a and b are chosen so that $a(t)$ is close enough to $x(t)$ and $b(t)$ is close enough to $y(t)$ for e to satisfy $e_1(t) > y_1(t)$, $y_2(t) > e_2(t)$, $e_2(t) > x_2(t)$ and $e_1(t) < x_1(t)$. If $x_1(t) < y_1(t)$ and $x_2(t) > y_2(t)$ we can again assume that a and b are chosen so that $a(t)$ is close enough to $x(t)$ and $b(t)$ is close enough to $y(t)$ for e to satisfy $e_1(t) < y_1(t)$, $y_2(t) < e_2(t)$, $e_2(t) < x_2(t)$, and $e_1(t) > x_1(t)$. Now choose $z \in Z(e)$ such that $\forall \, t \in I \cup J$: (i) $x_1(t) > z_1(t) > y_1(t)$ and $x_2(t) < z_2(t) < y_2(t)$ hold if $x_1(t) > y_1(t)$ and (ii) $x_1(t) < z_1(t) < y_1(t)$ and $x_2(t) > z_2(t) > y_2(t)$ hold if $x_1(t) < y_1(t)$. We can choose $z(t) \ll e(t)$ close enough to $e(t)$ for the equalities to hold. In fact, we can choose a, b, e, and z simultaneously so that $a(t)$ and $b(t)$ are close enough to $x(t)$ and $y(t)$ respectively for $x(t)$, $y(t)$, and $z(t)$ to lie on an SCDS curve for each $t \in I \cup J$.

Suppose that we have $(x, y) \in \bigcap_{i \in I} Ap(i) \bigcap_{j \in J} - Ap(j)$. Then by Lemma 10.7 there exists $p' \in W$ such that $(x, z), (z, y) \in \bigcap_{i \in I} Ap'(i) \bigcap_{j \in J} - Ap'(j)$ and $p'(t) = p(t) \; \forall \, t \notin I \cup J$. We have $z \in Z(a) \cap Z(b)$. And there exists $\varepsilon \in B_0$ such that $Z(\varepsilon) \subset Z(a) \cap Z(b) \cap Z(\omega)$. By Lemma 4 the pair (I, J) is decisive for $f \,|\, Z(\varepsilon)$, and thus (I, J) is decisive for $f \,|\, Z(a)$ and $f \,|\, Z(b)$ because $Z(\varepsilon)$ contains free triples. Therefore, $(x, z) \in Af(p')$ because x and z belong to $Z(a)$ and (I, J) is decisive for $f \,|\, Z(a)$. And $(y, z) \in Af(p')$ because y and z belong to $Z(b)$ and (I, J) is decisive for $f \,|\, Z(b)$. Then we have $(x, y) \in Af(p')$ because $Af(p')$ is transitive, and thus $(x, y) \in Af(p)$ by Arrow's independence axiom. Therefore, (I, J) is decisive for $f \,|\, \{x, y\}$.

Now we show that the members of I have veto power and the members of J have inverse veto power for $f \,|\, \{x, y\}$. Suppose that $1 \in I$ and $(x, y) \in Ap(1) \cap - Af(p)$. Choose $p' \in W$ so that $(x, z), (z, y) \in Ap'(1)$ and $(z, y) \in \bigcap_{i \in I} Ap(i) \bigcap_{j \in J} - Ap(j)$ and

$p'(t) \cap \{x,y\}^2 = p(t) \cap \{x,y\}^2 \forall t \in T$. Then $(y,x) \in Af(p')$ by Arrow's independence axiom and $(z,y) \in Af(p')$ by decisiveness of (I,J) over $f \mid Z(b)$. Then we have $(z,x) \in Af(p')$ because $Af(p')$ is transitive. This contradicts the veto power of 1 for $f \mid Z(a)$. Therefore, each member of I has veto power for $f \mid \{x,y\}$.

Suppose that $1 \in J$ and $(x,y) \in Ap(1) \cap Af(p)$. Choose $p' \in W$ so that $(x,z),(z,y) \in Ap(1)$ and $(y,z) \in \cap_{i \in I} Ap(i) \cap_{j \in J} - Ap(j)$ and $p'(t) \cap \{x,y\}^2 = p(t) \cap \{x,y\}^2 \forall t \in T$. Then $(x,y) \in Af(p')$ by Arrow's independence axiom and $(y,z) \in Af(p')$ by decisiveness of (I,J) over $f \mid Z(a)$. Then we have $(x,z) \in Af(p')$, contradicting the inverse veto power of 1 for $f \mid Z(b)$. Therefore, each member of J has inverse veto power for $f \mid \{x,y\}$. ∎

6. Theorem. If $f: W(B_0) \to Q(B_0)$ satisfies strict non-imposition and Arrow's independence axiom then f is strictly oligarchical.

Proof. As in the proof of Lemma 5, let (I,J) denote the oligarchy underlying $f \mid Z(\omega)$. We know that $I \cup J$ is finite (Lemma 9.12). Suppose that $(x,y) \in \cap_{i \in I} Ap(i) \cap_{j \in J} - Ap(j)$. Choose $t \in I$. Suppose that $x(t) \geq y(t)$ holds. There exist a finite number n of vectors $z^1(t), z^2(t), \ldots, z^n(t)$ in $\mathbf{E}^2_{++}$ such that:

6a $z^1(t) = x(t)$ and $z^n(t) = y(t)$.
6b $z_1^i(t) + z_2^i(t) > z_1^{i+1}(t) + z_2^{i+1}(t)$ for $i = 1,2,\ldots,n-1$.
6c $z^i(t) \geq z^{i+1}(t)$ is false for $i = 1,2,\ldots,n-1$.
6d $z^i(t) \leq z^{i+1}(t)$ is false for $i = 1,2,\ldots,n-1$.

The vectors $z^i(t)$ can be found by connecting $x(t)$ and $y(t)$ by a straight line $L(t)$ and then drawing n equally spaced lines with slope -1; have the first line go through $x(t)$ and the last one go through $y(t)$. By choosing $z^i(t)$ and $z^{i+1}(t)$ on opposite sides of $L(t)$ and far enough from $L(t)$ we can ensure that 6c and 6d hold. Obviously, 6b will hold if $z^i(t)$ is on a higher negatively sloped line than $z^{i+1}(t)$. If $t \in I$ and neither $x(t) \geq y(t)$ nor $y(t) \geq x(t)$ holds then $L(t)$ has a negative slope and we can obviously find $z^1(t)$, $z^2(t), \ldots, z^n(t)$ on $L(t)$ such that 6a holds and $(z^i, z^{i+1}) \in AR$ for some $R \in W_t$ and $i = 1,2,\ldots,n = 1$. Suppose that $t \in J$ and $y(t) \geq x(t)$. Then we can find a finite number n of vectors $z^1(t)$, $z^2(t), \ldots, z^n(t)$ in $\mathbf{E}^2_{++}$ such that 6a, 6c, and 6d hold, and 6b is satisfied with the reverse inequality. If $t \in J$ and neither $x(t) \geq y(t)$ nor $y(t) \geq x(t)$ holds then $L(t)$ has a negative slope and we can obviously find $z^1(t), z^2(t), \ldots, z^n(t)$ on $L(t)$ such that 6a holds and $(z^{i+1}, z^i) \in AR$ for some $R \in W_t$ and $i = 1, 2, \ldots, n - 1$. Because $I \cup J$ is finite we can have the same n for all $t \in I \cup J$.

Now, we wish to define $z^i(t)$ for $t \notin I \cup J$ so that the members $z^1, z^2, \ldots, z^n$ of the finite sequence are all in B_0 and there exists a profile p' in W such that $(z^i, z^{i+1}) \in \bigcap_{h \in I} Ap'(h) \bigcap_{j \in J} - Ap'(j)$ for $i = 1, 2, \ldots, n - 1$. Because $I \cup J$ is finite we can choose a finite n large enough to ensure that the vector $z^i(t)$ can be chosen so that for each $t \in I \cup J$ it is arbitrarily close to $x^i(t) \equiv [((n - i)/(n - 1)] x(t) + [(i - 1/(n - 1)] y(t)$. Because x and y belong to B_0 the allocation x^i belongs to B_0. Because B_0 is defined by strict inequalities the allocation z^i can be found in B_0 if n is large enough. (The larger we make n, the closer we can get $z^i(t)$ to $x^i(t)$.)

Now we can choose $p' \in W$ such that $(z^i, z^{i+1}) \in \bigcap_{h \in I} Ap(h) \bigcap_{j \in J} - Ap(j)$ for $i = 1, 2, \ldots, n - 1$ and $p'(t) = p(t) \, \forall t \notin I \cup J$. We have $z^i \in F_t(z^{i+1})$ for $t \in I \cup J$ and thus $(z^i, z^{i+1}) \in Af(p')$ by Lemma 5. Therefore, $(x, y) \in Af(p')$ by transitivity of $Af(p')$ and thus we have $(x, y) \in Af(p)$ by Arrow's independence axiom. Therefore, (I, J) is decisive over all pairs of allocations in B_0.

Now, suppose that we have $1 \in I$ and $(x, y) \in Ap(1) \cap - Af(p)$. Construct the sequence $z^1, z^2, \ldots, z^n$ as above. Choose $p' \in W$ so that $(x, z^2) \in Ap'(1)$ and $(z^i, z^{i+1}) \in \bigcap_{h i \in I} Ap'(h) \bigcap_{j \in J} -Ap'(j)$ for $i = 2, 3, \ldots, n - 1$. We have $(y, x) \in Af(p')$ by Arrow's independence axiom, and $(z^i, z^{i+1}) \in Af(p')$ for $i = 2, 3, \ldots, n - 1$ by decisiveness of (I, J). Therefore, (z^2, y) by transitivity of $Af(p')$ and hence $(z^2, x) \in Af(p')$ because (y, x) belongs to $Af(p')$. This contradicts the veto power of 1 over pairs $\{x, z^2\}$ such that $x \in F_t(z^2) \forall t \in I \cup J$ (Lemma 5). Therefore, 1 must have veto power for f. Similarly, the members of J have inverse veto power for f. ∎

We have assumed a starting-date $t = 1$ for notational convenience but it is clear that the arguments are all valid when time extends backward to minus infinity as well as forward to plus infinity. Now, we consider the case of fully transitive social preferences.

Choose $\alpha \in \mathbf{E}^T_{++}$ and β and define the SCDS curve $Y_t(\alpha, \beta) = \{x \in \mathbf{E}^2_{++} \mid \log x_1(t) + \alpha(t) \log x_2(t) = \beta(t)\}$. Set $Y(\alpha, \beta) = \Pi_{t \in T} Y_t(\alpha, b)$. Suppose that $f: W \to P(\Omega)$ satisfies Arrow's independence axiom, Ω is locally connected in the topology employed, and $f \mid \Omega_0$ is not constant. Lemma 10.9 shows that $f \mid Y(\alpha, \beta)$ is not constant for some choice of $\alpha \in \mathbf{E}^T_{++}$ and $\beta \in \mathbf{E}^T$.

The argument that $f \mid Y(\alpha,\beta)$ is actually authoritarian depends on finiteness of T but that is the only part of the proof of 10.9 that is not valid for any T. The proof of Lemma 10.9 also depends on the assumption that if x and y belong to Ω_0 then every neighbourhood of x contains a point z such that, for all $t \in T$, $x(t) \neq y(t)$ implies $x(t) \neq z(t) \neq y(t)$. This is a mild assumption on the topology and it will be implicit in the discussion to follow. If $f \mid Y(\alpha,\beta)$ is not constant then by Lemma 2 and Theorem 9.6, $f \mid Y(\alpha,\beta)$ is completely authoritarian for some α and β. For concreteness, assume that 1 dictates $f \mid Y(\alpha,\beta)$. Choose $\alpha' \in \mathbf{E}_{++}^{T}$ and $\beta' \in \mathbf{E}^{T}$ so that $Y(\alpha,\beta)$ and $Y(\alpha',\beta')$ meet, at z say. Define V_t by splitting both $Y_t(\alpha,\beta)$ and $Y_t(\alpha',\beta')$ into two pieces at z and stitching the highest left-hand and right-hand pieces to make the SCDS curve V_t. Then V_t itself is an SCDS curve and $W(V)$ is basic, for $V = \Pi_{t \in T} Y_t$. Choose $x \in Y(\alpha,\beta) \cap V$ and $y \in Y(\alpha',b') \cap V$ so that $x \neq z \neq y$ and $\{x,z\}$ and $\{y,z\}$ are free. Then 1 is a dictator for $f \mid V$ by Theorem 9.6 because $x,z \in Y(\alpha,\beta) \cap V$ and 1 dictates $f \mid \{x,z\}$. Therefore 1 dictates $f \mid Y(\alpha',\beta')$ because $z,y \in V \cap Y(\alpha',\beta')$ and 1 dictates $f \mid \{z,y\}$.

If $Y(\alpha,\beta) \cap Y(\alpha',\beta') = \varnothing$ choose α'',β'' such that $Y(\alpha'',\beta'')$ meets both $Y(\alpha,\beta)$ and $Y(\alpha',\beta')$. Then 1 dictates $f \mid Y(\alpha'',\beta'')$ by the above argument, and thus 1 dictates $f \mid Y(\alpha',\beta')$. We have proved the following: if f is not constant there is some individual who is a dictator for all $f \mid Y(\alpha,\beta)$ or else there is someone who is an inverse dictator for all $f \mid Y(\alpha,\beta)$.

Now we prove the infinite society counterpart to Theorem 10.13. The reader is reminded that the definition of a basic domain restricts the choice of a topology for the outcome space. The restriction is a mild one, however. In particular, we need not require continuity of social preference with respect to the product topology.

7. Theorem. Assume a topology for which Ω is locally connected and each product of SCDS curves is a connected Hausdorff space. If $f: W \to P(\Omega)$ is a non-constant social welfare function satisfying Arrow's independence axiom then f is completely authoritarian.

Proof. Suppose that f is a non-constant social welfare function satisfying Arrow's independence axiom. If $f \mid \Omega_0$ is constant then so is f by Lemma 10.4. Therefore $f \mid \Omega_0$ is not constant. Therefore, some $t \in T$ is a dictator (or an inverse dictator) for $f \mid Y(\alpha,\beta)$ and all $\alpha \in \mathbf{E}_{++}^{T}$ and $\beta \in \mathbf{E}^{T}$ as we have just demonstrated.

Suppose that t is a dictator for all $f \mid Y(\alpha,\beta)$. Choose arbitrary $x,y \in \Omega_0$ and $p \in W$. Suppose that (x,y) belongs to $Ap(t)$. Choose any $\alpha,\beta,\alpha',\beta'$ such that $x \in Y(\alpha,\beta)$ and $y \in Y(\alpha',\beta')$, and $\{x,z\}$ and $\{z,y\}$ are free for some $z \in Y(\alpha,\beta) \cap Y(\alpha',\beta')$, and $(x,z),(z,y) \in AR'$ for some $R' \in W_t$. Clearly, $z \in \Omega_0$. Define $p' \in W$ by setting $p'(i) = p(i)$ for all $i \neq t$ and $p'(t) = R'$. Then $(x,z),(z,y) \in Af(p')$ and hence $(x,y) \in Af(p')$ by transitivity. Therefore, $(x,y) \in Af(p)$ by independence and thus t is a dictator for $f \mid \Omega_0$.

Now, $W(\Omega_0)$ is obviously regular (8.9). Therefore, $f \mid \Omega_0$ is completely authoritarian by Proposition 8.10. Then $f(p) \cap \Omega_0^2 = p(t) \cap \Omega_0^2$ for all $p \in W$ and thus $f(p) = p(t)$ for all $p \in W$ by Lemma 10.4. That is, f is completely dictatorial.

If no individual is a dictator for all $f \mid Y(\alpha,\beta)$ then some individual is an inverse dictator for all $f \mid Y(\alpha,\beta)$. The mirror image of the above argument will show that this person is an inverse dictator for f and f is completely authoritatian. ∎

Now we can prove the corresponding theorem for the overlapping generations economy itself. Recall that $B = \{x \in \Omega \mid x_2(t-1) + x_1(t) \leq \omega_2(t-1) + \omega_1(t) \; \forall t \in T\}$ and $B_0 = \{x \in \Omega \mid (0,0) \leqslant x_2(t-1) + x_1(t) \leqslant \omega_2(t-1) + \omega_1(t) \; \forall t \in T\}$, where $\omega_2(0) = 0$.

8. Theorem. Assume that B is both connected and locally connected and that $B_0 \cap Y$ is connected for each product Y of SCDS curves. The social welfare function $f: W(B) \to P(B)$ satisfies Arrow's independence axiom if and only if it is constant or completely authoritarian.

Proof. (i) Obviously, f satisfies Arrow's independence axiom if it is constant or completely authoritarian.

(ii) Suppose that $f: W(B) \to P(B)$ satisfies Arrow's independence axiom. It is clear that the argument of Theorem 7 works for any product space $X = \Pi_{t \in T} X_t$ as long as each X_t has a non-empty interior in $\mathbf{E}_+^2$, $Y \cap X$ is connected in the chosen topology for each product Y of SCDS curves, and X is locally connected in the chosen topology. Recall that $Z(\omega) = \{y \in \Omega_0 \mid y(t) \leqslant \omega(t) \; \forall t \in T\}$ is a subset of B_0. Because $Z(\omega)$ is a product set $f \mid Z(\omega)$ is completely authoritarian. Let t be the dictator or inverse dictator for $f \mid Z(\omega)$. Then either $(\{t\},\varnothing)$ or $(\varnothing,\{t\})$ is an oligarchy for $f \mid Z(\omega)$. Then by Lemma 5 and Theorem 6, the oligarchy for $f \mid Z(\omega)$ is an oligarchy for $f \,|\, B_0$. That is, $f \,|\, B_0$ is either dictatorial or inversely dictatorial. Then $f \,|\, B_0$ is completely authoritarian by Proposition 8.10. Therefore, f itself is completely authoritarian by Lemma 10.4. ■

13   Implementation

IN this chapter the impossibility theorems of Chapter 10 are stated in terms of a mild incentive compatibility condition instead of Arrow's independence axiom. This substitution results in criteria whose value and appropriateness are more transparent. In order to introduce strategic considerations we begin with a *social choice correspondence* instead of a social welfare function.

1. Social choice correspondence. A social choice correspondence for outcome space X and domain D is a correspondence Φ on $D \times X$ that associates with each profile p in D and each subset Z of X a subset $\Phi(p,Z)$ of Z.

The definition allows $\Phi(p,Z)$ to be empty in some cases. The social choice criteria would have to require non-emptiness of $\Phi(p,Z)$ for cases that are of practical importance. To this end, let ψ denote the prescribed family of subsets Z of X for which $\Phi(p,Z)$ must contain at least one alternative for each $p \in D$. Of course, $\Phi(p,Z)$ is interpreted as the set of socially optimal outcomes in Z when individual preferences are specified by p. We will be more formal about this (Definition 4) but, essentially, by implementability of Φ we mean that for each candidate feasible set $Z \in \psi$ a mechanism can be found such that $\Phi(p,Z)$ is the set of equilibria generated by the mechanism for arbitrary profile p. If Φ is implementable in that sense we also say that Φ is incentive compatible. (We allow different feasible sets to be associated with different mechanisms. Indeed, it would be hard to decide formally if a mechanism defined for feasible set $\{a,b\}$ is identical to a mechanism defined for feasible set $\{x,y,z\}$.)

A social welfare function f on a domain D associates with each preference profile p in the domain a social preference relation $f(p)$ over the universal outcome space X. It generates a social choice correspondence $\max f$ on $D \times X$. For each $p \in D$ and $Z \subset X$, let $\max f(p,Z)$ denote the set of $f(p)$-maximal alternatives x in Z. That is $\max f(p,Z) = \{x \in Z \mid (y,x) \in Af(p) \text{ implies } y \notin Z\}$. This replaces the slightly more cumbersome notation $\max f(p)(Z)$ that results when we apply the definition of Chapter 1. (See 1.11.) In this chapter the social choice correspondence $\max f$ is required to satisfy a simple condition that is necessary for implementability in terms of co-operative or non-co-operative equilibrium. The condition, which we call Plott's independence axiom, is implicit in all of the standard papers on the implementation of social choice correspondences; see, for example, Maskin (1977), Palfrey and Srivistava (1991), Williams (1986), Abreu and Sen (1990, 1991), Moore and Repullo (1988), Saijo (1988), Jackson (1989), Herrero and Srivistava (1990). Although the condition is very mild it implies that the social welfare function f is either constant or dictatorial if $f(p)$ itself is required to be a continuous preorder on the space X of allocations of two or more private goods and D is the domain of classical economic preferences.

It is well known that implementation in dominant strategies is extremely demanding. As we saw in Chapter 7, if the social choice correspondence is single valued then only trivial or dictatorial rules are so implementable. Even if the choice set can be multivalued, implementation in dominant strategies is possible only under extreme and unpalatable restrictions on the social choice correspondence. This is proved in Kelly (1977), Pattanaik (1978), and Ferejohn, *et al.* (1982). In this chapter we merely require implementation at Nash equilibrium points, or at subgame-perfect Nash equilibria, or with respect to some other Nash refinement, or even with respect to extended Nash equilibria in which some coalitions have power–strong Nash equilibrium for example (Maskin, 1979), or coalition-proof Nash equilibrium (Bernheim *et al.*, 1987). (It should be pointed out that Jackson (1989) proves that the additional requirement of bounded message spaces makes even these definitions of implementation very restrictive.) The Plott independence axiom is implied by any of the Nash refinements and by Nash equilibrium itself. It is implicit in Maskin (1977), Saijo (1988), Williams (1986), and Matsushima (1988) on Nash implementation, in Moore and Repullo (1988) and Abreu and Sen (1990) on subgame-perfect Nash implementation, in Palfrey and Srivistava (1991) on implementation in undominated strategies, in Abreu and Sen (1987) on virtual implementation, and in Herrero and Srivistava (1990) on implementation via backward induction. In each of these treatments implementation is possible for a very wide range of social choice correspondences. Nevertheless, the Plott independence axiom implies that the *social welfare function* is either authoritarian or constant, and the Plott independence axiom is only a mild necessary condition! Obviously, dictatorship or constancy is sufficient for implementation, but an inversely dictatorial social welfare function is not incentive-compatible. Therefore, a social welfare function is incentive-compatible if and only if it is constant or completely dictatorial, as Theorem 8 below will establish. (The fact that a constant social welfare function is the only alternative to dictatorship is parallel to the Gibbard–Satterthwaite result which proves that a non-manipulable voting scheme is either dictatorial or it ignores all but two of the outcomes.)

This chapter also proves an incentive-compatibility result to parallel the oligarchy theorem (10.15). If social preference is merely required to be quasitransitive and the social welfare function f satisfies strict non-imposition then implementability implies that f is properly oligarchical.

Now we begin presenting a formal statement of our incentive-compatibility criterion. In Plott (1976 and 1986) the following condition on the social choice correspondence Φ is shown to be implied by a wide variety of implementation requirements. Note that the condition bears on the family ψ of candidate feasible sets as well as on Φ.

2. Plott's independence axiom. The social choice correspondence Φ satisfies Plott's independence axiom with respect to ψ if for all $Z \in \psi$ and all $p, p' \in D$ such that $p(t) \cap Z^2 = p'(t) \cap Z^2$ holds for all $t \in T$ we have $\Phi(p, Z) = \Phi(p', Z)$.

Here is the rationale for Plott's axiom. If, agent by agent, individual preferences over Z are the same under p' as under p then any equilibrium for p' is an equilibrium for p and vice versa. By definition, implementation requires that $\Phi(p,Z)$ is the set of outcomes associated with equilibria at p and thus $\Phi(p',Z) = \Phi(p,Z)$. To be more precise, assume that Φ is a social choice correspondence for outcome space X and domain D. Given Z, let ϕ denote the sub-correspondence $\Phi(\cdot, Z)$ on D. Suppose that (S,g) is a mechanism implementing ϕ. If x belongs to $\phi(p)$ then there is some $s^* \in S$ such that $g(s^*) = x$ and s^* is an equilibrium with respect to p. *Assuming* that infeasible alternatives— i.e. outcomes not in Z—play no role in guiding the mechanism (S,g) to an equilibrium and *assuming* that individual preferences involving infeasible alternatives have no effect on behaviour, then s^* is also an equilibrium for any $p' \in D$ such that $p'(t) \cap Z^2 = p(t) \cap Z^2$ for all $t \in T$. Because (S,g) implements ϕ and s^* is an equilibrium for p' we must have $g(s^*) \in \phi(p')$. Therefore, $\phi(p) = \phi(p')$ if individual preferences over Z are identical in the two situations, regardless of our definition of equilibrium, as long as the assumptions about individual behaviour and the mechanism's adjustment process are appropriate.

We will be concerned with the incentive compatibility of a given social welfare function f, by which we mean implementation of max$f$, and we conclude that implementability considerations imply that max$f$ satisfies Plott's independence axiom. That is, for all $p, p' \in D$ and all $Z \in \psi$, max$f(p,Z) = $ max$f(p', Z)$ if $p(t) \cap Z^2 = p'(t) \cap Z^2$ for all $t \in T$. In fact, we will restrict attention to ψ_m, the family of m-element subsets of X, where m is a fixed integer larger than unity. The proof assumes a finite society and thus the results of this chapter apply to the finite society case only. Proposition 3 establishes that when Plott's independence axiom is in force for some finite integer m exceeding unity then for every $x,y \in \Omega$ and all $p, p' \in W$ we have $f(p) \cap \{x,y\}^2 = f(p') \cap \{x,y\}^2$ if $p(t) \cap \{x, y\}^2 = p'(t) \cap \{x,y\}^2 \; \forall \, t \in T$. This means that f satisfies Arrow's independence axiom if max f satisfies Plott's independence axiom with respect to ψ_m for some $m \geq 2$.

If max$f$ satisfies Plott's independence axiom with respect to ψ_2 then f obviously satisfies Arrow's independence axiom. However, we prove this implication for arbitrary $m \geq 2$, and the conclusion is far from obvious if m exceeds 2. Note in particular that Plott's axiom does not immediately imply that $f(p)$ and $f(p')$ agree over Z when p and p' agree over Z; we merely have agreement between the sets of maximal elements, max$f(p,Z)$ and max$f(p',Z)$. We will be concerned with the family ψ_m of m-element subsets of X for a simple reason. Computational economy — information-processing considerations, etc. — demands that the feasible sets Z be finite. If they are too small, however, the approximation to the 'true' continuum-feasible set is poor. We will think of m as a large finite number, but the results are valid for any finite $m \geq 2$. (The rationale for assuming discreteness when addressing implementation questions is discussed in Hurwicz and Marschak (1985) and in Marschak (1987). What we have called Plott's independence axiom is referred to as Independence of Infeasible Alternatives in Grether and Plott (1982).)

Now we prove a simple proposition: Plott's independence axiom implies Arrow's independence axiom if $\psi = \psi_m, m \geq 2, X = \Omega$, and $D = W$.

3. Proposition. Assume a finite society T, a finite integer $m \geq 2$, and a social welfare function $f: W \to Q(\Omega)$. Then max$f$ satisfies Plott's independence axiom with respect to ψ_m if and only if f satisfies Arrow's independence axiom.

Proof. (i) Obviously, Arrow's independence axiom implies Plott's for any set ψ.

(ii) Assume that $m \geq 2$ is a finite integer, that f maps W into $Q(\Omega)$, and that max$f$ satisfies Plott's independence axiom with respect to ψ_m. Suppose that x and y belong to Ω, p and p' belong to $W, p(t) \cap \{x,y\}^2 = p'(t) \cap \{x,y\}^2$ for all $t \in T$ and $(x,y) \in Af(p)$ but $(y,x) \in f(p')$. Set $I = \{t \in T \mid (x,y) \in Ap(t)\}$ and $J = \{t \in T \mid (y,x) \in Ap(t)\}$. Because both individual and social preference relations are continuous there are neighbourhoods $N_t(y)$, $N(y)$, $M_t(y)$ of y such that $\{x\} \times N_t(y) \subset Ap(t) \, \forall \, t \in I$, $\{x\} \times N_t(y) \subset -Ap(t) \, \forall t \in J$, and $\{x\} \times N(y) \subset Af(p)$; and also $\{x\} \times M_t(y) \subset Ap'(t) \, \forall t \in I$, $\{x\} \times M_t(y) \subset -Ap'(t) \, \forall t \in J$. Set $U(y) = \cap_{i \in I} N_i(y) \cap_{j \in J} N_j(y) \cap N(y) \cap_{i \in I} M_i(y) \cap_{j \in J} M_j(y)$. Choose $m - 2$ alternatives $z^1, z^2, \ldots, z^{m-2}$ in $U(y)$ such that $z^1(t) \gg z^2(t) \gg \ldots z^{m-2}(t) \gg y(t), \, \forall \, t$. Set $Z = \{x, y, z^1, z^2, \ldots, z^{m-2}\}$. Selfishness and monotonicity imply $p(t) \cap (Z - \{x\})^2 = p'(t) \cap (Z - \{x\})^2, \, \forall t \in T$. If $(x,y) \in Ap(t) \cup -Ap(t)$ we have $p(t) \cap \{x,z^i\}^2 = p'(t) \cap \{x,z^i\}^2$ for $1 \leq i \leq m - 2$ by choice of $U(y)$ and the fact that $U(y)$ contains z^i. If $(x,y) \in Sp(t)$ we have $(z^i, x) \in Ap(t) \cap Ap'(t)$ by transitivity of individual preference and the fact that (z^i, y) belongs to $Ap(t) \cap Ap'(t)$. Therefore, $p(t) \cap Z^2 = p'(t) \cap Z^2, \, \forall t \in T$. Then max$f(p,Z) = $ max$f(p',Z)$ by Plott's axiom. We have $(x,v) \in Af(p) \, \forall \, v \in U(y)$ and thus max$f(p,Z) = \{x\}$. When Z is finite, $Af(p')$ is transitive, and max$f(p',Z) = \{x\}$ we cannot have $(y,x) \in f(p')$ as we now prove.

Claim. If R is a complete and quasitransitive relation on X and Y is a finite subset of X then max$R(Y) \neq \{x\}$ if $(y,x) \in R$ for some $y \in Y - \{x\}$.

Proof of claim. Obviously, max$R(Y) \neq \{x\}$ if $x \notin Y$. Suppose that x and y belong to Y, $(y,x) \in R$, and $x \neq y$. If $Y = \{x,y\}$ the claim is certainly true. Suppose that the claim is true whenever Y has n elements. Now, consider an arbitrary set Y with $n + 1$ elements ($n \geq 2$). Choose some $z \in Y - \{x,y\}$. Set $Z = Y - \{z\}$. If $x \notin \maxR(Z)$ then $x \notin \maxR(Y)$. If $x \in \maxR(Z)$ then $\maxR(Z) \neq \{x\}$ by the induction hypothesis and thus there is some $w \in \maxR(Z) - \{x\}$. If $w \in \maxR(Y)$ we are finished. If $w \notin \maxR(Y)$ then $(z,w) \in AR$. If $z \in \maxR(Y)$ we are finished. Otherwise there is some $v \in Y$ such that $(v,z) \in AR$. Then $(v,w) \in AR$ by transitivity of AR. But $v \in Z$ contradicting $w \in \maxR(Z)$.

We have proved the claim and thus we have established that $f|\Omega_0$ satisfies Arrow's independence axiom. ∎

Proposition 3 can be compared with Blau (1971). Assuming $m \geq 2$, an unrestricted domain, and a set of X of at least $m + 2$ alternatives Blau proves that Arrow's independence axiom is equivalent to the requirement that $f(p) \cap Z^2 = f(p') \cap Z^2$ whenever $Z \in \psi_m$ and $p(t) \cap Z^2 = p'(t) \cap Z^2 \forall t \in T$.

Now we use Proposition 3 to extend Theorems 10.13 and 10.15 to statements about the implementability of social welfare functions on W. Before defining implementability formally we outline the proofs of the two theorems. The critical assumption is that infeasible alternatives play no role in the operation of the mechanism and have no bearing on individual behaviour. Then Plott's independence condition is implicit if we suppose that $\max f$ is implementable over ψ_m. First, we can put Proposition 3 together with Theorem 10.13 to declare that $f : W \rightarrow P(\Omega)$ is implementable over ψ_m only if f is constant or completely dictatorial or completely inversely dictatorial. Completely dictatorial social welfare functions can be implemented in terms of dominant strategies. Inversely dictatorial social welfare functions can be dismissed out of hand on ethical grounds. In the context of resource allocation with private goods they are even worse than dictatorial rules. The allocation that assigns the zero commodity vector to every individual is always socially 'optimal' according to an inversely dictatorial rule. Even though inverse dictatorship is absurd we will take the trouble to eliminate it on incentive compatibility grounds; we will prove that $f : W \rightarrow P(\Omega)$ is implementable over ψ_m only if it is constant or completely dictatorial (Theorem 8). Similarly, if $f : W \rightarrow Q(\Omega)$ is completely oligarchical and implementable there can be no member of the oligarchy whose preferences are inverted before they are incorporated into the group preference relation (Theorem 10).

To justify the rubric 'theorem' we formally define implementability of f over ψ_m. Recall the definition of a mechanism (7.5).

4. Implementability. The social welfare function $f : W \to P(\Omega)$ is implementable with respect to ψ if for each $Z \in \psi$ there is a mechanism γ_Z such that for each $p \in W$ the set $\gamma_Z(p)$ of equilibria of the induced game coincides with $\max f(p,Z)$, the set of $f(p)$-maximal elements in Z.

The definition is incomplete because we have not defined an equilibrium. To prove the theorem all that we need to say about the equilibrium concept is that it has the following property: if $p, p' \in W$ and $p(t) \cap Z^2 = p'(t) \cap Z^2 \forall t \in T$ then $\gamma_Z(p) = \gamma_Z(p')$. This obviously implies Plott's independence axiom, assuming that $\gamma_Z(p) = \max f(p,Z)$ holds for all p and Z.

5. Theorem. Assume a finite society T and a finite integer $m \geq 2$. If the social welfare function $f : W \to P(\Omega)$ is implementable with respect to ψ_m then it is constant or completely authoritarian.

Proof. Implementability implies Plott's independence axiom. Therefore, f satisfies Arrow's independence axiom by Proposition 3. Therefore, f is constant or completely dictatorial or completely inversely dictatorial by Theorem 10.13. ∎

It is obvious that this result can be sharpened; an inversely dictatorial social welfare function surely cannot be implemented. We make this formal by sharpening the definition of implementability, and we do that by giving more structure to the notion of equilibrium. We now define *Nash implementation* which is implementation in the sense of Definition 4 when the equilibrium employed has the following Nash property.

6. The Nash property. If s^* is an equilibrium at profile p of a mechanism defined for the set Z, and $x^* \in Z$ is the outcome generated by s^*, and $(x,x^*) \in Ap(t)$ for all $x \in Z - \{x^*\}$ and $t \in T$, then s^* is an equilibrium for every profile in the domain.

In words, if s^* is an equilibrium even when everyone prefers every other available alternative to the one associated with s^*, then s^* will still be an equilibrium when the outcome associated with it rises in the preference ordering of some individuals. Nash equilibrium has this property, as does strong Nash equilibrium.

7. Nash implementability. The social welfare function $f: W \to P(\Omega)$ is Nash implementable over ψ if for each $Z \in \psi$ there is a mechanism γ_Z such that for each $p \in W$ the set $\gamma_Z(p)$ of equilibria of the induced game coincides with $\max f(p,Z)$ for some equilibrium concept that has the Nash property.

8. Theorem. Assume a finite society T and a finite integer $m \geq 2$. If the social welfare function $f: W \to P(\Omega)$ is Nash implementable with respect to ψ_m then it is constant or completely dictatorial.

Proof. According to Theorem 5 we just have to rule out inverse dictatorship. Suppose that t is an inverse dictator for f. Choose x^1, $x^2, \ldots, x^m \in \Omega$ such that $\log x_1^i(t) + \log x_2^i(t) = 0$ for all $t \in T$ and $x_1^i(t) < x_1^{i+1}(t)$ for $i = 1, 2, \ldots, m-1$ and all $t \in T$. Then we can find $p, p' \in W$ satisfying $(x^1, x) \in Ap(i) \cap -Ap'(i)$ for all $i \in T$ and $x \in Z - \{x^1\}$. Then $\max f(p', Z) = \{x^1\}$ so there is some equilibrium s^* of the game associated with p' such that the outcome generated by s^* is x^1. Then s^* is also an equilibrium with respect to p by the Nash property. But $x^1 \notin \max f(p,Z)$. Therefore, f is not Nash implementable. ∎

Compare Theorem 8 to the Gibbard–Satterthwaite theorem. The hypothesis of the former is much more demanding in one dimension: the social choice correspondence Φ must be rationalizable. For each profile p there is a binary relation $f(p)$ *on the universal outcome set X* such that $\Phi(p,Z)$ is the set of $f(p)$-maximal members of Z for *each admissible Z*. In other words, $\Phi = \max f$ for some social welfare function f. Our hypothesis is much weaker in every other respect. We do not assume that $\Phi(p,Z)$ is a singleton. Dominant strategies are not required. And the domain of individual preferences is severely restricted, although it must be noted that Barberà and Peleg (1990) prove the Gibbard–Satterthwaite theorem for the domain of continuous individual preorders with respect to a meaningful topology, and Moreno and Walker (1991) and Zhou (1991) prove weak versions of the theorem for the domain of economic preferences. (The hypothesis of the Moreno–Walker theorem is very close to that of the Gibbard–Satterthwaite theorem.)

Compare our incentive-compatibility theorem to Arrow's seminal result (5.5). Arrow's independence axiom, the Pareto criterion, and the free-triple property are eliminated from the hypothesis. The new theorem requires only the mild incentive-compatibility condition, which is weaker than Arrow's independence axiom in general, and a continuity requirement that is natural in an economic setting. Full transitivity of social preference is the one common assumption. Dictatorship is the only possibility under Arrow's hypothesis, and in our framework the only additional social welfare functions to qualify are the constant ones.

Compare Theorem 5 to Moore and Repullo (1988) or Abreu and Sen (1990). Both papers prove that, *given* the feasible set Z, almost any social choice subcorrespondence $\Phi(\cdot, Z)$ is implementable with respect to subgame-perfect Nash equilibrium. (The subcorrespondence must satisfy a very modest monotonicity requirement.) Virtually all correspondences defined over economic environments are so implementable. However, this chapter assumes that the feasible set Z is variable *and* that $\Phi(p, Z)$ is connected to $\Phi(p, Z')$ by means of a social ordering on X; consequently, a strong impossibility theorem emerges. The existence of a social welfare function causes the problem, a striking confirmation of Arrow's original insight.

The corresponding results for quasitransitive social preference is not as sharp because it requires the assumption of strict nonimposition. The first theorem follows directly from Proposition 3 and Theorem 10.15

9. Theorem. Assume that $m \geq 2$. If the social welfare function $f: W(\Omega_0) \to Q(\Omega_0)$ is implementable and satisfies strict nonimposition then it is completely oligarchical.

10. Theorem. Assume that $m \geq 2$. The social welfare function $f: W(\Omega_0) \to Q(\Omega_0)$ is Nash implementable and satisfies strict nonimposition if and only if it is completely and properly oligarchical.

Proof. (i) Suppose that f is properly and completely oligarchical. Then it meets the Maskin criterion for implementability at Nash equilibrium points if the oligarchy contains at least two members (Maskin 1977; Saijo 1988). If the oligarchy is a singleton then the social welfare function is dictatorial and hence obviously Nash implementable.

(ii) Implementability implies Plott's independence axiom. Therefore, f satisfies Arrow's independence axiom by Proposition 3. Therefore, f is completely oligarchical by Theorem 10.15. Let (I, J) be the underlying oligarchy.

Let $Z = \{x^1, x^2, \ldots, x^m\}$ be the m-element subset of Ω_0 specified in the proof of Theorem 8. Choose two profiles p and p'

in W such that $(x, x^m) \in Ap(t)$ for all $t \in T$ and $x \in Z - \{x^m\}$, and $(x^1, x^m) \in \cap_{i \in I} Ap'(i) \cap_{j \in J} -Ap'(j)$. If J, the set of individuals whose preferences are inverted, is non-empty we have $x^m \in \max f$ (p, Z). Alternative x is socially preferred to x^m only if everyone in I prefers x to x^m *and* everyone in J prefers x^m to x, and there is no such alternative in Z. Then $x^m \in \max f(p, Z)$ so there is some equilibrium strategy configuration s^* for p such that x^m is the outcome precipitated by s^*. But x^m is everyone's least-preferred outcome under p. Therefore, s^* is an equilibrium for every profile by the Nash property, and in particular for p'. That is, x^m belongs to $\gamma_Z(p')$ and thus $x^m \in \max f(p', Z)$, a contradiction because everyone in I prefers x^1 to x^m everyone in J prefers x^m to x^1. Therefore, J must be empty. ∎

Theorems 5 and 9 contrast sharply with the many positive results on implementation. These positive results show that a wide range of social choice correspondences can be implemented at Nash (or subgame-perfect Nash, etc.) equilibrium points. Theorem 5, on the other hand, claims that only constant or authoritarian social welfare functions can be implemented, and Theorem 9 proves that within the family of social welfare functions satisfying strict non-imposition only the oligarchical rules are implementable. There is no paradox once it is realized that Theorems 5 and 9 are statements about social welfare functions as much as they are statements about implementation. Implementation considerations are used to justify Plott's independence axiom which in turn implies Arrow's original independence axiom. With respect to Theorem 5, all the work really consists in showing that the Arrow axiom by itself — without the Pareto criterion or even non-imposition — implies that the social welfare function is constant or dictatorial or inversely dictatorial if social preference is continuous and fully transitive. *Given* a set Z the subcorrespondence $\max f(\cdot, Z)$ is implementable for many other social welfare functions. But if f is fixed and Z is variable then even mild implementability considerations place severe restrictions on f and these do not surface in the standard implementation literature, which employs a single feasible set. Alternatively, one can say that the standard implementation literature does not assume that the social choice correspondence can be rationalized by a social welfare function.

The proof of the proposition linking implementability to Arrow's independence axiom depends crucially on the assumption that the feasible sets are finite. We can define a simple utilitarian social welfare function that is implementable with respect to a special family of 'Edgeworth box' feasible sets. Let ψ_E be the family of subsets $E(\delta)$ *of* Ω such that x belongs to $E(\delta)$ if and only if $\Sigma_{t\in T}x(t) \leq \delta(1,1, \ldots ,1)$, where δ is an arbitrary positive number. To define $f(p)$, the social preorder for $p \in W$, let u_t be the continuous utility representation of $p(t)$ that satisfies $u_t(\lambda,\lambda, \ldots ,\lambda) = \lambda$ for all real $\lambda \geq 0$. Because u_t is unique we can define a social welfare function f by letting $f(p)$ be the preorder generated by $\Sigma_{t\in T}u_t$. For each $Z \in \psi_E$ the projection correspondence $\max f(\cdot,Z)$ satisfies the Moore–Repullo or Abreu–Sen sufficient condition for subgame implementation. (Assume that T has at least three members.) Clearly, f is neither constant nor dictatorial. The family ψ_E of feasible sets is rather contrived; each member is an Edgeworth *square*. As a result, the utility representations restricted to $Z \in \psi_E$ are identical for two individual preorders that agree over Z. Whether there exist reasonable social welfare functions that are implementable with respect to a realistic family ψ of connected feasible sets — or even Edgeworth boxes — is an important open question.

14 Conclusion

THE most striking theorem in this book reveals that the Arrow independence axiom is an obstacle to efficiency–equity trade-offs. Any step away from authoritarianism, however small, takes us outside of the family of collective choice rules that are responsive to individual preferences, however slight the degree of responsiveness. In plainer words, if the social welfare function is not authoritarian then it is constant! And not all authoritarian social welfare functions are responsive. Dictatorship is the only collective choice rule satisfying Arrow's independence axiom that also responds positively to individual preference. One other assumption is vital to this impossibility theorem: social preference is represented by a continuous binary relation that is also transitive. But why continuity, and why transitivity?

Continuity is important for practical reasons. Empirically, it is impossible to distinguish allocation x from y if the Euclidean distance between them is sufficiently small. Therefore, if x is socially preferred to another allocation z then y should be as well, and each must be socially preferred to allocations sufficiently close to z. Practically speaking, continuity is inevitable and it is difficult to see why the theoretical model should not reflect reality in this respect.

Transitivity is a separate issue. We have acknowledged that intransitivity of the social indifference relation does not prevent the identification of maximal allocations in compact sets, and have considered the possibility of defining a satisfactory social welfare function satisfying Arrow's independence axiom and continuity and quasitransitivity of social preference. But we have not taken the investigation far enough. Although we have substituted strict non-imposition for the stronger Pareto criterion, the former is still unacceptably strong.

The Pareto criterion insists on the social ranking of x above y when every individual prefers x to y. Strict non-imposition opens the door to a social welfare function that ranks x above y only when everyone receives substantially more benefit under x than under y.

The strict non-imposition condition avoids specifying what it means for an individual to receive substantially more benefit from x by merely requiring that x rank above y socially for *some* configuration of preferences (unless some individual receives more of every good under y or there is some other a priori restriction on *individual* preferences over $\{x,y\}$). But consider the following example involving private goods only. Allocation x gives every individual one hundred times more of every good than y, except for commodity c which each person consumes slightly more of under y. One can find classical — even linear — indifference maps that rank y above x for each individual. The non-imposition condition implies that there is at least one preference profile under which y ranks above x socially. We might not object much to a social welfare function that always ranks x above y if it is attractive from an equity standpoint, if it respects efficiency when it really counts, and if it yields continuous and quasitransitive social preferences. This question is not settled in the preceding pages, although there are reasons for pessimism. First, if we select two vital commodities and hold the consumption of all other goods constant we are led to the consideration of a partial social welfare function f_z, where z represents the given allocation of the goods whose consumption is held constant. It is reasonable to impose strict non-imposition on f_z because the two commodities are vital, and it would be unacceptable if f_z always ranked x above y if y gave everyone one hundred times more of good 1 and only slightly less of good 2. But it would also be unacceptable if f_z were oligarchical. (If the oligarchy were the entire society then equity would play no role in the distribution of the two goods, and otherwise at least one person would be left out of the distribution of the two goods in every situation.) The second reason for pessimism is that if we assume non-imposition and Arrow's independence axiom then the only social welfare function that generates continuous quasiorders and is sensitive to the preferences of every individual in a non-perverse way is the Pareto aggregation rule, which ranks x above y socially if and only if everyone strictly prefers x to y. This rule does not stand in opposition to equity, but it is completely uninformative on matters of distributional justice. Again we see that relaxation of the Pareto, or efficiency, criterion does not open the door to equity. (Campbell and Nagahisa (1991) provides a more cogent version of this argument.)

We have used an exchange economy to make the point that dictatorship is egregiously inequitable in the case of private goods and selfish and monotonic individual preferences; all goods will be confiscated and transferred to the dictator. Some people object to the use of an exchange economy, pointing out that in the real world less output would be produced as more and more goods were transferred to the dictator. This is true, of course, but it does not vitiate the use of an exchange economy as a proving ground. A good test of a social welfare function's intrinsic equity properties is its performance in an exchange economy, and both the dictatorial and the inversely dictatorial rules fail dismally.

Arrow's independence axiom is the driving force behind the impossibility theorems in this book, and it remains a controversial condition. We have shown how a mild incentive-compatibility condition can be substituted without vitiating any of the main theorems. There remain five ingredients that could arguably be weakened. The first is strict non-imposition. The second is incentive compatibility. The incentive-compatibility condition that we have employed implicitly assumes that individuals base their strategic behaviour on ordinal preference data only. Thirdly, the assumption of quasitransitivity could be further weakened to acyclicity. Nagahisa (1991) has made a start. Fourthly, the requirement that social choice be based on a binary relation could itself be relaxed, although Grether and Plott (1982) have shown that this will not necessarily destroy the impossibility theorem virus. And there is a sound reason for requiring a social preference relation. In practice, alternatives often present themselves in pairs. Whenever a community considers a departure from the status quo or an amendment to a motion it is deliberating on a pair of social states. Finally, the requirement that the entire space X of options has to be socially ordered at the outset could be relaxed. However, Chapter 12 demonstrates that the impossibility theorems go through when attention is confined to the feasible set of allocations, even though it is not a product set. However, consideration of non-ordinal individual preference information and the investigation of acyclicity without non-imposition, and even of non-binary social choice in economic environments, are important next steps.

References

ABREU, D., and SEN, ARUNAVA (1990), 'Subgame Perfect Implementation: A Necessary and almost Sufficient condition', *Journal of Economic Theory* 50: 285–99.
—— —— (1991), 'Virtual Implementation in Nash Equilibrium', *Econometrica* 59: 997–1022.
ALIPRANTIS, C. D., and BROWN, D. J. (1983), 'Equilibria in Markets with a Riesz Space of Commodities', *Journal of Mathematical Economics* 11: 189–207.
ALLEN, B. (1987), 'Smooth Preferences and the Approximate Expected Utility Hypothesis', *Journal of Economic Theory* 41: 340–55.
ARROW, K. J. (1951), *Social Choice and Individual Values*, 1st edn. (New York: Wiley).
—— (1963), *Social Choice and Individual Values*, 2nd edn. (New York: Wiley).
AUERBACH, A. J., and KOTLIKOFF, L. J. (1987), *Dynamic Fiscal Policy* (Cambridge: Cambridge University Press).
BAIGENT, N., and HUANG, P. (1990), 'Topological Social Choice: Reply to Le Breton and Uriarte', *Social Choice and Welfare* 7: 141–6.
BARBERÀ, S. (1983a), 'Strategy-Proofness and Pivotal Voters: A Direct Proof of the Gibbard–Satterthwaite Theorem', *International Economic Review* 24: 413–18.
—— (1983b), 'Pivotal Voters: A Simple Proof of Arrow's Theorem', in P. K. Pattanaik and M. Salles (eds.), *Social Choice and Welfare* (Amsterdam: North-Holland), 31–5.
—— and PELEG, B. (1990), 'Strategy-Proof Voting Schemes with Continuous Preferences', *Social Choice and Welfare* 7: 31–8.
—— SONNENSCHEIN, H., and ZHOU, L. (1991), 'Voting by Committees', *Econometrica* 59: 595–610.
BERGSTROM, T. C. (1975), 'Maximal Elements of Acyclic Relations on Compact Sets', *Journal of Economic Theory* 10: 403–4.
BERLIANT, M. (1984), 'An Equilibrium Existence Result for an Economy with Land', *Journal of Mathematical Economics* 12.
BERNHEIM, B. D., PELEG, B., and WHINSTON, M. D. (1987), 'Coalition-proof Nash Equilibria: Concepts', *Journal of Economic Theory* 42: 1–12.
BEWLEY, T. F. (1972), 'Existence of Equilibria in Economies with

178 *References*

Infinitely Many Commodities', *Journal of Economic Theory* 4: 514–40.

BINMORE, K. (1976), 'Social Choice and Parties', *Review of Economic Studies* 43: 459–64.

BLACK, D. (1948), 'On the Rationale of Group Decision-making', *Journal of Political Economy* 56: 23–4.

—— (1958), *The Theory of Committees and Elections* (London: Cambridge University Press).

BLAU, J.H. (1957), The Existence of Social Welfare Functions', *Econometrica* 25: 302–13.

—— (1971), 'Arrow's Theorem with Weak Independence', *Economica* 38: 413–20.

BLIN, J.-M., and SATTERTHWAITE, M.A. (1976), 'Strategy-proofness and Single-peakedness', *Public Choice* 26: 51–8.

—— —— (1978), 'Individual Decisions and Group Decisions: The Fundamental Differences', *Journal of Public Economics* 10: 247–67.

BORDER, K.C. (1983), 'Social Welfare Functions for Economic Environments with and without the Pareto Principle', *Journal of Economic Theory* 29: 205–16.

BORDES, G., and M. LE BRETON (1989), 'Arrovian Theorems with Private Alternatives Domains and Selfish Individuals', *Journal of Economic Theory* 47: 257–82.

—— —— (1990), 'Arrovian Theorems for Economic Domains: The Case when there are Simultaneously Private and Public Goods', *Social Choice and Welfare* 7: 1–18.

—— and N. Tideman (1991), 'Independence of Irrelevant Alternatives in the Theory of Voting', *Theory and Decision* 30: 163–86.

BOURBAKI, N. (1966), *General Topology*, ii (Reading, Mass.: Addison-Wesley).

BROWN, D.J., and LEWIS, L.M. (1981), 'Myopic Economic Agents', *Econometrica* 49: 359–68.

CAMPBELL, D.E. (1985), 'Impossibility Theorems for Infinite Horizon Planning', *Social Choice and Welfare* 2: 283–93.

—— (1989a), 'Arrow's Theorem for Economic Environments and Effective Social Preferences', *Social Choice and Welfare* 6: 325–9.

—— (1989b), 'Wilson's Theorem for Economic Environments and Continuous Social Preferences', *Social Choice and Welfare* 6: 315–23.

—— (1990a), 'Can Equity be Purchased at the Expense of Efficiency? An Axiomatic Enquiry', *Journal of Economic Theory* 51: 32–47.

—— (1990b), 'Intergenerational Social Choice without the Pareto Principle', *Journal of Economic Theory* 50: 414–23.

—— (1991a), 'Transitive Social Choice in Economic Environments', *International Economic Review*, forthcoming.

CAMPBELL, D. E. (1991*b*), 'Implementation of Social Welfare Functions', *International Economic Review*, forthcoming.

—— (1991*c*), 'Overlapping Generations' Economies and Efficiency–Equity Trade-Offs', Working Paper, College of William and Mary, Williamsburg, Va.

—— (1991*d*), 'Public Goods and Arrovian Social Choice', *Social Choice and Welfare*, forthcoming.

—— (1992), 'Quasitransitive Intergenerational Choice for Economic Environments', *Journal of Mathematical Economics*, forthcoming.

—— and Nagahisa (1991), 'A Simple Axiomatization of Pareto Optimality', Working Paper, College of William and Mary, Williamsburg, Va.

—— and WALKER, M. (1990), 'Maximal Elements of Weakly Continuous Relations', *Journal of Economic Theory* 50: 459–64.

CHICHILNISKY, G. (1980), 'Social Choice and the Topology of Spaces of Preferences', *Advances in Mathematics* 37: 165–76.

—— (1982*a*), 'Social Choice and Game Theory: Recent Results with a Topological Approach', in P. K. Pattanaik and M. Salles (eds.), *Social Choice and Welfare* (Amsterdam: North-Holland), 79–102.

—— (1982*b*), 'Social Aggregation Rules and Continuity', *Quarterly Journal of Economics* 97: 337–52.

CONDORCET, M. DE (1785), *Essai sur l'Application de l'Analyse à la Probabilité des Décisions Rendues à la Pluralité des Voix* (Paris).

CRAWLEY, P., and DILWORTH, R. P. (1973), *Algebraic Theory of Lattices* (Englewood Cliffs, NJ.: Prentice-Hall).

DASGUPTA, P., HAMMOND, P. J., and MASKIN, E. S. (1979), 'Implementation of Social Choice Rules', *Review of Economic Studies* 46: 181–216.

DIAMOND, P. (1965), 'The Evaluation of Infinite Consumption Streams', *Econometrica* 33: 170–7.

DIERKER, E. (1971), 'Equilibrium Analysis of Exchange Economies with Indivisible Commodities', *Econometrica* 39: 997–1008.

DONALDSON, D., and ROEMER, J. (1987), 'Social Choice in Economic Environments with Dimensional Variation', *Social Choice and Welfare* 4: 253–76.

—— and WEYMARK, J. (1988), 'Social Choice in Economic Environments', *Journal of Economic Theory* 46: 291–308.

DUGUNDJI, J. (1966), *Topology* (Boston, Mass.: Allyn & Bacon).

EDGEWORTH, F. Y. (1881), *Mathematical Psychics* (London: Kegan Paul).

FEREJOHN, J. A., and FISHBURN, P. C. (1979), 'Representation of Binary Decision Rules by Generalized Decisiveness Structures', *Journal of Economic Theory* 21: 28–45.

—— —— and McKELVEY, R. D. (1982), 'Implementation of Democratic Social Choice functions', *Review of Economic Studies* 49: 439–46.

— GRETHER, D.M., MATTHEWS, S.A., and PACKEL, E.W. (1980), 'Continuous-valued binary decision procedures, *Review of Economic Studies* 47: 787-96.

— and PACKEL, E.W. (1983), 'Continuous Social Decision Procedures', *Mathematical Social Sciences* 6: 64-73.

FISHBURN, P.C. (1970), 'Arrow's Impossibility Theorem: Concise Proof and Infinite Voters', *J. Economic Theory* 2: 103-6.

FOUNTAIN, J., and SUZUMURA, K. (1982), 'Collective Choice rules without the Pareto Principle', *International Economic Review* 23: 299-308.

GIBBARD, A. (1969), 'Social Choice and the Arrow Condition', unpublished MS.

— (1973), 'Manipulation of Voting Schemes: A General Result', *Econometrica* 40: 587-602.

GRETHER, D.M., and PLOTT, C.R. (1982), 'Nonbinary Social Choice: An Impossibility Theorem', *Review of Economic Studies* 49: 143-9.

GUHA, A.S. (1972), 'Neutrality, Monotonicity, and the Right of Veto', *Econometrica* 40: 821-6.

HERRERO, M.J., and SRIVISTAVA, S. (1990), 'Implementation via Backward Induction', Graduate School of Industrial Administration, Carnegie-Mellon University, Pittsburgh.

HURWICZ, L., and MARSCHAK, T. (1985), 'Discrete Allocation Mechanisms: Dimensional Requirements for Resource-Allocation Mechanisms when Desired Outcomes are Unbounded', *Journal of Complexity* 1: 264-303.

JACKSON, M.O. (1989), 'Implementation in Undominated Strategies: A Look at Bounded Mechanisms', Center for Mathematical Studies in Economics and Management Science, Northwestern University Discussion Paper 833.

KALAI E., and MEGIDDO, N. (1980), 'Path Independent Choices', *Econometrica* 48: 781-4.

— and MULLER, E. (1977), 'Characterization of Domains Admitting Nondictatorial Social Welfare Functions and Nonmanipulable Voting procedures', *Journal of Economic Theory* 16: 457-69.

— MULLER, E., and SATTERTHWAITE, E. (1979), 'Social Welfare Functions when Preferences are Convex, Strictly Monotonic, and Continuous', *Public Choice* 34: 87-97.

— and RITZ, Z. (1980), 'Characterization of the Private Alternatives Domains admitting Arrow Social Welfare Functions', *Journal of Economic Theory* 22: 23-36.

KELLEY, J.L. (1955), *General Topology* (New York: Van Nostrand).

KELLY, J.S. (1971), 'The Continuous Representation of a Preference Ordering', *Econometrica* 39: 593-7.

KELLY, J.S. (1977), 'Strategy-Proofness and Social Choice Functions without Singlevaluedness', *Econometrica* 45: 439–46.
— (1978), *Arrow Impossibility Theorems* (New York: Academic Press).
— (1987), *Social Choice Theory: An Introduction* (Berlin: Springer-Verlag).
KELSEY, D. (1988), 'What is Responsible for the "Paretian Epidemic"?' *Social Choice and Welfare* 5: 303–6.
KIRMAN, A.P. (1987), 'Vilfredo Pareto', in J. Eatwell, M. Milgate, and P. Newman (eds.), *The New Palgrave: A Dictionary of Economics* (Macmillan: New York).
— and SONDERMANN, D. (1972), 'Arrow's Theorem, Many Agents, and Invisible Dictators', *Journal of Economic Theory* 5: 267–77.
LE BRETON, M., and URIARTE, J.R. (1990), 'On the Robustness of the Impossibility Result in the Topological Approach to Social Choice', *Social Choice and Welfare* 7: 131–40.
MACHINA, M.J. (1982), '"Expected Utility" Analysis without the Independence Axiom', *Econometrica* 50: 277–323.
MCMANUS, M. (1982), 'Some Properties of Topological Social Choice Functions', *Review of Economic Studies* 49: 447–60.
MARSCHAK, T. (1987), 'Price versus Direct Revelation: Informational Judgements for Finite Mechanisms', in T. Groves, R. Radner, and S. Reiter (eds.), *Information, Incentives and Economic Mechanisms: Essays in Honour of Leonid Hurwicz* (Oxford: Blackwell), 132–79.
MAS-COLELL, A. (1977), 'Indivisible Commodities and General Equilibrium Theory', *Journal of Economic Theory* 16: 443–56.
— and SONNENSCHEIN, H. (1972), 'General Possibility Theorems for Group Decisions', *Rev. Econ. Studies* 39: 185–92.
MASKIN, E.S. (1975a), 'Social Welfare Functions on Restricted Domains', mimeo, Harvard University, Cambridge, Mass.
— (1975b), 'Arrow Social Welfare Functions on Restricted Domains: The Two-Person Case', mimeo, Harvard University, Cambridge, Mass.
— (1977), 'Nash Equilibrium and Welfare Optimality', mimeo, MIT, Cambridge, Mass.
— (1979), 'Implementation and Strong Nash-Equilibrium', in J.-J. Laffont (ed.), *Aggregation and Revelation of Preference* (Amsterdam: North-Holland), 433–9.
MATSUSHIMA, H. (1988), 'A New Approach to the Implementation Problem', *Journal of Economic Theory* 45: 128–44.
MONJARDET, B. (1983), 'On the Use of Ultrafilters in Social Choice Theory', in P.K. Pattanaik and M. Salles (eds.), *Social Choice and Welfare* (Amsterdam: North-Holland), 73–8.
MOORE, J., and REPULLO, R.L. (1988), 'Subgame Perfect Implementation', *Econometrica* 56: 1191–220.
MORENO, D., and WALKER, M. (1991), 'Nonmanipulable Voting

Schemes when Participants' Interests are Decomposable', *Social Choice and Welfare* 8: 221–234.

MURAKAMI, Y. (1961), 'A Note on the General Possibility Theorems of the Social Welfare Function', *Econometrica* 29: 244–6.

NAGAHISA, R. (1991), 'Acyclic and Continuous Social Choice in T_1-connected Spaces: Including Its Application to Economic Environments', *Social Choice and Welfare* 8: 319–32.

PALFREY, T., and SRIVISTAVA, S. (1991), 'Nash Implementation using Undominated Strategies', *Econometrica* 59: 497–502.

PARETO, V. (1896), *Cours d' Économie Politique*, 2 vols. (Lausanne: Librarie de l'Université).

—— (1909), *Manuel d'Économie Politique*, 2nd edn. (Paris: Girard).

PATTANAIK, P.K. (1978), *Strategy and Group Choice* (Amsterdam: North-Holland).

PLOTT, C.R. (1976), 'Axiomatic Social Choice Theory: An Overview and Interpretation', *American Journal of Political Science* 20: 511–96.

—— (1986), 'A Relationship between Independence of Irrelevant Alternatives, Game Theory, and Implementability', mimeo, California Institute of Technology, Pasadena, Calif.

REDEKOP, J. (1991), 'Social Welfare Functions on Restricted Domains', *Journal of Economic Theory* 53: 396–427.

REPULLO, R.L. (1986), 'On the Revelation Principle under Complete and Incomplete Information', in K. Binmore and P. Dasgupta (eds.), *Economic Organizations as Games* (Oxford: Basil Blackwell), 179–95.

SAIJO, T. (1988), 'Strategy Space Reduction in Maskin's Theorem: Sufficient Conditions for Nash Implementation', *Econometrica* 56: 693–700.

SAMUELSON, P.A. (1958), 'An Exact Consumption Loan Model of Interest, with or without the Social Contrivance of Money', *Journal of Political Economy* 66: 467–82.

SATTERTHWAITE, M. (1975), 'Strategy-Proofness and Arrow's Conditions: Existence and Correspondence Theorems for Voting Procedures and Social Welfare Functions', *Journal of Economic Theory* 10: 187–217.

SCHMEIDLER, D., and SONNENSCHEIN, H.F. (1978), 'Two Proofs of the Gibbard–Satterthwaite Theorem on the Possibility of a Strategy-Proof Social Choice Function', in H.W. Gottinger and W. Leinfeller (eds.), *Decision Theory and Social Ethics: Issues in Social Choice* (Dordrecht: Reidel), 227–34.

SCHMITZ, N. (1977), 'A Further Note on Arrow's Impossibility Theorem', *Journal of Mathematical Economics* 4: 189–96.

SEN, A.K. (1970), *Collective Choice and Social Welfare* (San Francisco: Holden-Day).

—— (1986a), 'Information and Invariance in Normative Choice', in W.P. Heller, R.M. Starr, and D.A. Starrett (eds.), *Essays in Honor of K.J.*

Arrow: Vol. 1 Social Choice and Public Decision-Making, (Cambridge University Press, Cambridge), 29–55.

SEN, A.K. (1986*b*), 'Social Choice Theory', in K.J. Arrow and M.D. Intriligator (eds.), *Handbook of Mathematical Economics*, iii (Amsterdam: North-Holland), 1073–1181.

—— and PATTANAIK, P.K. (1969), 'Necessary and Sufficient Conditions for Rational Choice under Majority Decision', *Journal of Economic Theory* 1: 178–202.

TIAN, G. (1990), 'Necessary and Sufficient Conditions for the Existence of Maximal Elements of Preference Relations', Dept. of Economics, Texas A & M, College Station, Tex.

—— and ZHOU, J. (1990), 'Transfer Continuities, Generalizations of the Weierstrass and Maximum Theorems — A Characterization Approach', Dept. of Economics, Texas A & M, College Station, Tex.

WALKER, M. (1977), 'On the Existence of Maximal Elements', *Journal of Economic Theory* 16: 470–4.

WILLIAMS, S. (1986), 'Realization and Nash Implementation: Two Aspects of Mechanism Design', *Econometrica* 54: 139–52.

WILSON, R.B. (1972), 'Social Choice Theory without the Pareto Principle', *Journal of Economic Theory* 5: 478–86.

ZHOU, L. (1991), 'Impossibility of Strategy-Proof Mechanisms in Economies with Pure Public Goods', *Review of Economic Studies* 58: 107–19.

Subject Index

Name Index